Inventing Iron Man

IRON

E. Paul Zehr

THE POSSIBILITY OF
A HUMAN MACHINE

MAN

THE JOHNS HOPKINS UNIVERSITY PRESS | BALTIMORE

© 2011 E. Paul Zehr
All rights reserved. Published 2011
Printed in the United States of America on acid-free paper
9 8 7 6 5 4 3 2 1

The Johns Hopkins University Press
2715 North Charles Street
Baltimore, Maryland 21218-4363
www.press.jhu.edu

Library of Congress Cataloging-in-Publication Data

Zehr, E. Paul.
   Inventing iron man : the possibility of a human machine / E. Paul Zehr.
      p. cm.
   Includes bibliographical references and index.
   ISBN-13: 978-1-4214-0226-0 (hardcover : alk. paper)
   ISBN-10: 1-4214-0226-2 (hardcover : alk. paper)
  1. Cyborgs.   2. Androids.   3. Human physiology.   4. Human-machine
systems.   I. Title.
   TJ211.Z42 2011
   629.8—dc22   2011000177

A catalog record for this book is available from the British Library.

*Special discounts are available for bulk purchases of this book. For more
information, please contact Special Sales at 410-516-6936 or specialsales@
press.jhu.edu.*

The Johns Hopkins University Press uses environmentally friendly book
materials, including recycled text paper that is composed of at least 30
percent post-consumer waste, whenever possible.

To the memory of my mom, Marlene Mary Zehr (1935–2010), who passed away during the time I was working on this book. She always encouraged and supported me . . . and bought me my first comic books. Here's looking at you, Mom.

The suit of Iron Man and I are one.
—Robert Downey Jr. as Tony Stark in *Iron Man 2*

Learning is the only thing the mind never exhausts, never fears, and never regrets.
—Leonardo da Vinci (1452–1519)

Contents

Foreword, by Warren Ellis   ix

Preface: The Stark Reality of Robotics   xi

PART I. IT'S MORE THAN SKIN DEEP
        Tony learns to live inside a
        suit of iron

CHAPTER 1. Origins of the Iron Knight:
*Bionics, Robotic Armor, and Anthropomorphic
Suits*   3

CHAPTER 2. Building the Body with Biology:
When the Man of Metal Needs to Muscle In   16

CHAPTER 3. Accessing the Brain of the Armored Avenger:
Can We Connect the Cranium to a Computer?   37

The First Decades of Iron: "He Lives! He Walks! He
Conquers!"   65

PART II. USE IT AND LOSE IT
              Will time tarnish the Golden
              Avenger?

CHAPTER 4. Multitasking and the Metal Man:
    How Much Can Iron Man's Mind Manage?   69

CHAPTER 5. Softening Up a Superhero:
    *Why the Man with a Suit of Iron Could Get a
    Jelly Belly*   80

CHAPTER 6. Brain Drain:
    Will Tony's Gray Matter Give Way?   94

    The Next Decades of Iron: "I Can Envision the Future"   111

PART III. ARMORED AVENGER IN ACTION
              If we build it, what will come?

CHAPTER 7. Trials and Tribulations of the Tin Man:
    *What Happens When the Human Machine
    Breaks Down*   115

CHAPTER 8. Visions of the Vitruvian Man:
    Is Invention Really Only One Part Inspiration?   131

CHAPTER 9. Deal or No Deal?
    Could Iron Man Exist?   154

Appendix: Ten Momentous Moments of the Metal Man   181

Bibliography   185

Index   197

Foreword

Some years ago, I was approached by Marvel Comics editor-in-chief Joe Quesada to write new adventures of an early Marvel property that was still in continuous publication but, as these old-time superheroes tend to get, needed a new coat of modern paint. Iron Man was one of those characters that Marvel was having trouble getting a hold of. No hook to hang him on to capture the new century's light. I said to Joe, "He's the test pilot for the future. That's the whole thing. Flying away from himself, trying to bring the future on. Of all Stan Lee's ideas of the early sixties, this is the one that can and should reinvent itself annually, to keep pace with the stormfront of the future."

Of all pop culture's heroes of the past 50 years, Tony Stark is the one corporate-owned character who is absolutely designed to face the future. His armor is a reflective surface in which we can consider our era's own reaction to technological concerns. He began in a time when a weapons designer could still be a hero and when some people could still fantasize that the administering of savage beatings to Communists was the work of good men. At roughly the same time as Tony Stark was stomping through comics pages in his original tank-like armor, the U.S. military was testing the similarly massive and clunky Hardiman powered exoskeleton.

Today, Tony Stark is a bootstrapping "compassionate capitalist" attempting to bring free energy to the masses, and the Iron Man lives inside his bones as nine pints of colloidal technology. Even now we work, in the real world, on synthetic muscles, contact lenses with computer displays, cochlear implants, and fleets of nanoscale devices to sail our bloodstreams and keep us healthy. Tony Stark is the fictive ghost of our own cyborg tendencies, always a few years and at least one impossible idea ahead of us. Pop culture's test pilot for the future.

This wonderful book lays out the schema for that notion in energetic, eclectic detail. Starting from the only true way to see the Iron Man—as a prosthesis—the author uses the science fiction of Iron Man in its most effective way, as a tool with which to examine the present and past. From the Nyctalope and battlefield prosthetics of the eighteenth century to cutting-edge cortical implantation, the Iron Man is used as I and so many other writers in the world of comics and film have used it: as a metaphor. We simply hoped the use of the metaphor would intrigue and illuminate. The author achieves both effortlessly and has written a book that educates and delights. I hope you enjoy it as much as I did.

Warren Ellis
Southend, England

Preface

THE STARK REALITY OF ROBOTICS

Where is the line drawn? Between man and
machine? Where does her humanity end?
—Reflections on Tony Stark's assistant Pepper Potts
using the Iron Man armor, World's Most Wanted #1,
"Shipbreaking" (Invincible Iron Man #8, 2009)

I didn't realize at the time—that a shell of
iron—is hollow.
—Tony Stark on his childhood, "Dust to Dust"
(Iron Man #286, 1992)

Smashing through walls, flying through the air like a human jet, and controlling an amazingly complicated robotic suit of armor seemingly by mere thought. Oh, and being practically indestructible. These are things associated with the Marvel Comics character Iron Man. The full title description for his comic book is actually "The Invincible Iron Man," which is kind of a giveaway about the powers he is supposed to possess. Being invincible is a pretty tall order, though. Basically Iron Man is a really, really smart guy (OK—I give in, he is a genius) in a super high-tech suit of armor. It seems pretty clear that we humans have been able to develop some extraordinary technology. We certainly have the ability to control powerful machines and fly through the air (and beyond) with rockets and jets. We also have some pretty fancy armored suits for protection in outer space and in the deep dark reaches of inner space, the sea. But the ability to really put them all together at once is the key to having a real Iron Man. The central focus of this book is exploring just that issue. Is it possible to have seamless biological control of an armored robotic suit? And, if it is possible, what does it really mean for how our bodies function and for our future as human beings?

One day I was speaking with my daughters (then ages 6 and 9) about the focus of this book. I told them I was going to explore the background of the possibility of Iron Man. Since they didn't yet know about Marvel's character of Iron Man, I showed them some images in an Invincible Iron Man comic book from the 1970s. Their basic response was this: "He wears a big suit. I wouldn't want to be him; it looks too hot." I told them that the suit is "air-conditioned." While that answer satisfied my kids that day and may deal with the real but superficial problem of overheating both human and machine, it does not address at all the totally unsuperficial problem of how that big suit could actually be controlled by the human inside it. Exploring this problem and all the related problems that spring from it is the real focus of *Inventing Iron Man: The Possibility of a Human Machine.*

In many ways, this book carries on from my previous one, *Becoming Batman: The Possibility of a Superhero.* That book was about examining the reality behind whether the self-made man Bruce Wayne could become the ultimately sculpted Batman through physical training. This book is about the self-invented man Tony Stark becoming Iron Man through the application of robotics. However, as we shall learn, a lot of training and adaptations are needed to actually master that iron suit. The iconic characters I have studied have several things in common. Batman is also a superhero lacking in actual superpowers. He first appeared in Detective Comics; the first appearance of Iron Man wasn't in his own comic book either. Instead, his first story was in Tales of Suspense #39 from March 1963. As with so many Marvel characters, this story was the brainchild of creative genius and writer Stan Lee with synthetic contributions from scripter Larry Lieber and artists Don Heck and Jack Kirby. Just to make sure readers could not fail to note how powerful Iron Man was supposed to be, the cover of Tales of Suspense #39 actually says "Who? or what, is the newest, most breath-taking, most sensational super-hero of all . . . ? Iron Man! He lives! He walks! He conquers!" In our journey examining the possibility of Iron Man, we will talk about the living and the walking parts. I am a bit opposed to gratuitous conquering so we won't really get into that bit. In homage to the true comic book writing style, I will emphasize that by throwing down an exclamation point!

Iron Man is the character that emerges when millionaire industrialist Tony Stark (for those not in the know, his actual full name is Anthony Edward Stark) puts on a mechanized suit of armor that he custom designed and built. The basic origin story for Iron Man—like all comic characters—has been revised, revisited, and re-created over the years. The key part that has been maintained in all alternate origins is that Tony the industrialist gets captured and kidnapped by bad guys. They know (as does the whole world because Tony Stark really is a genius and brilliant inventor and head of a huge international conglomerate) that he can build all kinds of devices.

The actual story title for Iron Man's introduction in 1963 was "Iron Man Is Born!" and in this tale the military industrialist Tony Stark has been designing and selling weapons to the U.S. government for use against the communist guerillas in South Vietnam. While on a "site visit" to the jungles of Vietnam to see his technology in action, Tony trips a booby trap bomb. Shrapnel from this bomb gets lodged in his chest very near to his heart. In order to get help and a surgery to save his life, Tony agrees to help create a new weapon for Wong-Chu, who is the main communist terror warlord. However, he plans to trick and double-cross the villain with the help of another kidnapped scientist, physicist Professor Yinsen. The two men build a chest plate that creates a magnetic field which then acts to hold the shrapnel in a kind of stasis. This was really nicely shown in the 2008 Marvel Studios' film *Iron Man*. If you recall the scene in the desert cave in which Tony, played by Robert Downey Jr., wakes up to find a 12-volt car battery connected to terminals coming out of his chest, you get the idea. Well, from this chest plate Tony and Yinsen develop a full-body and fully articulated mechanized suit of armor. This is the double-cross part, by the way, and gives birth to Iron Man as a super hero to fight crime.

At this juncture, Tony Stark moves away from being the capitalist solely concerned with profit from making munitions to becoming the iron-garbed superhero and founding member of the Avengers—Iron Man. A key implication from this origin tale is that Tony Stark must always wear the chest plate to keep the shrapnel from moving into his heart and killing him. This was shown in a panel in Tales of Suspense #40 from 1963 in a story entitled "Iron Man vs. Gargantus." Tony is mulling over how sad he feels that he couldn't go swimming with a girlfriend (Jeanne), "She probably thought I was trying to avoid her, but I couldn't go swimming! I can never appear anywhere bare-chested because I constantly wear this iron chest plate. Just as other men plug in their electric shavers for their morning or evening shave, I must constantly charge up this plate which gives continued life to my heart!" When he finally plugs his chest plate into the outlet meant for the shaving razor, he exclaims "Ah! Electrical energy is pouring back! Now I can continue living . . . to help humanity as Iron Man!"

I am going to admit right up front that Iron Man plays a bit fast and loose with valid concepts of physics and energetics. All right, all right, those of you "in the know" realize that is really an understatement. I spoke to my colleague Jim Kakalios, the friendly neighborhood physics professor and author of *The Physics of Superheroes* (you should read this book—it is great) about this very issue. He pointed out that "energy storage in batteries has dramatically lagged behind information storage. If batteries had followed the Moore's Law that describes the increase in density of transistors on integrated circuits, with a doubling in capacity every two years, then a battery that would discharge in one hour in 1970 would last for over a century today. Ultimately, if we don't want to wear licensed nuclear power packs on our backs, we are limited to chemical processes to run our suit of high-tech armor, and in that case we must either sacrifice weight or lifetime." And on this point I concur with him completely. The energetic needs of Iron Man outstrip what we can provide currently. We cannot really power up to use repulsor rays and so on. However, what we are going to explore is just how much of the Iron Man character is based on a realistic extension of concepts in neuroscience, robotics, biomedical engineering, and kinesiology that we have today.

But, while it might seem perfectly realistic to you that suits of armor could be powered and worn rather like clothes, you likely haven't thought much about the real science behind creating ways for biological creatures like human beings to connect with artificial creations like bionic limbs. In the comic books, imagining this very thing was how Stan Lee and others created a superhero, albeit one quite different from the granddaddy of superheroes, Superman. Iron Man is a mere human being. As highlighted by Andy Mangels in his book *Iron Man: Beneath the Armor*, famous Iron Man writer David Michelinie has said that Iron Man is "a super hero with no superpowers. Any abilities he has are abilities that he makes, that he imagines and then invents. Prime among those, of course is his amazing suit of electronic armor. Without the armor, he's just a man. A man with a huge brain and a few billion dollars, but still just a human being . . . That makes him a lot more interesting than many heroes, as well as making him easier for the average reader to identify with. He could be you or me, if we had the money and inventiveness. And the courage. And the willpower."

Inventing Iron Man is divided into three parts related to different aspects of how humans can interface with technology. In Part I we start "skin deep" and then explore areas under the skin such as muscles and nerves. We also look at what might go on top of the skin in examining the concept of controllable suits of supporting armor by using Iron Man as a primary example and other attempts at creating prosthetic extensions of the human body as secondary examples. This exploration also includes discussing the way in which the body works normally so we can better appreciate the effect of layering technology over top of our biological machines. We also consider this from the perspective of something that you wear (or drive!) to survey the reality about what is needed to make connections between biological beings and machines. It really is largely all in our heads—which is to say our brains. We will talk about the possibility of controlling things like a computer or a robotic arm by measuring electrical activity in the brain and spinal cord. This is the area of neuroscience known as "brain-computer interface"—the literal connection between activity in the nervous system and actual machines.

Part II describes the long-term effects of interfacing with the kind of technology in the Iron Man suit. Our bodies adapt to the stresses that they experience, and interfacing with technology is a biological problem of stress adaptation. This kind of interface removes some stresses that are normally present in the body but also adds a few new ones. What are the limits within which our bodies can borrow, blend, and become one with artificial technology and, perhaps more importantly, how does this alter the body itself?

Part III looks at the good and the bad about Tony Stark the man and what he brings to the suit. Tony wrestles with demons—in a bottle and otherwise. But his drive and creativity allow him to constantly reinvent himself and his suit to changing circumstances. In this way, he has a kinship to creative geniuses the likes of Leonardo da Vinci and modern-day inventors such as Yoshiyuki Sankai and

Yves Rossy. If you don't know who these men are, you will by the time you have finished the book! The chapter also tackles the issue of what kinds of problems you might experience as an iron-suited superhero. What are some of the practical aspects of being Iron Man?

Inventing Iron Man explains the science behind and limitations of the extent to which human beings can control and interface with computers, machines, and robots. Because Iron Man is a normal human being inside a high-tech suit of armor, it is always assumed that anybody, well practically anybody, could just slap on the gear and be ready to go. This is not so, as you will read in this book. A lot of specialized learning and adaptations in the body of the Iron Man armor "user" would be needed. Just exactly what those adaptations are (and they aren't all good) you will find out later. For those of you who aren't that familiar with Iron Man, you will learn here a bit about a Marvel Comics icon and the science behind linking humans to machines. For those readers very familiar with Tony Stark and Iron Man, well, I have a few surprises ahead for you. To find out exactly what I mean, please keep on turning the pages and read along as together we probe the possibility of inventing Iron Man—the possibility of a human machine.

I remain inspired by my two main scientific mentors, Digby Sale at McMaster University and Richard Stein at the University of Alberta. They both kindled my interest in neuroscience, and I thank them for lighting and fanning that spark into a flame. I must also point out the accidental inspiration provided by Dan Ferris of the University of Michigan who, I think, planted a seed by showing Iron Man images in a conference talk on robotic exoskeletons given at Key Biscayne, Florida some years ago.

I conducted many interviews during the writing of this book, and I am indebted to those who agreed to speak and correspond with me. Yves (the "Jet Man") Rossy, Yoshiyuki Sankai and Fumi Takeuchi of Cyberdyne Inc., Phil Nuytten of Nuytco, Jon Wolpaw at Wadsworth Center, Doug Weber at the University of Pittsburgh, Max Donelan at Simon Fraser University, David Williams (formerly of the Canadian Space Agency, now of McMaster University), and David Wolf and Robert Frost of the National Aeronautics and Space Agency all corresponded with me at various times. I thank them very much for their time.

I remain impressed by the level of professionalism and competence at the Johns Hopkins University Press. I thank Vince Burke for his help throughout the entire process of proposal to publication, Michele T. Callaghan for her truly outstanding skills as copy editor, and Kathy Alexander for her tireless and effective work as publicist.

I also thank all the readers of my first book, *Becoming Batman: The Possibility of a Superhero*, who have so kindly provided feedback on how they enjoyed it. You helped sustain me in writing this book.

Last, thank you to Jordan, Andi, and Lori for helping to keep me grounded.

PART I IT'S MORE THAN SKIN DEEP

Tony learns to live inside a suit of iron

Origins of the Iron Knight

BIONICS, ROBOTIC ARMOR, AND ANTHROPOMORPHIC SUITS

> I thought, well, if a guy had a suit of armor, but it was a modern suit of armor—not like years ago in the days of King Arthur—and what if that suit of armor made him as strong as any Super Hero? I wasn't thinking robot at all: I was thinking of armor, a man wearing twentieth-century armor that would give him great power.
> —Comic book icon Stan Lee on his inspiration for creating Iron Man, in *Iron Man: Beneath the Armor* by Andy Mangels

> Iron Man is one of those comics where you have very few purists who have attached themselves to particular story lines. In the case of Iron Man it's the myth of Iron Man . . . it's the suit . . . it's what the suit could do.
> —Jon Favreau, director of the 2008 movie *Iron Man*

The prototypical British heavy metal band Black Sabbath rang in 1970 with Tony Iommi's immortal guitar riff (heavily distorted courtesy of Laney amps) and words (heavily distorted courtesy of Ozzy Osbourne's vocal cords and, um, distinct manner of speech),

forever giving us the phrase "I . . . am . . . Iron Man." That musical
Iron Man was cast as a villain who has a vision of a future apocalypse. But who is Iron Man the comic book icon? Let's be honest: first
impressions are often mostly visual. And at first glance, the defining
visual characteristic of Iron Man is his iron armor. The main attraction and defining characteristic for Iron Man really is skin deep. The
Marvel Comics character Iron Man certainly represents the most
well-known comic character to wear a suit of armor like it is his own
skin. Colossus from the Uncanny X-Men also has an iron skin, except in his case it is literal in that his skin actually changes into iron!
John Henry Irons from DC Comics also has an iron body, with capabilities that rival Superman. And even Batman once used an oversized exoskeleton to help defeat Superman in Frank Miller's *The
Dark Knight Returns*. Yet none of these well-known and lesser known
heroes is the combination pilot, soldier, police officer, deep-sea diver,
and flying human that Tony Stark is.

Many buzzwords could be used to describe the themes we are
going to explore in this book, including "bionics" and "cybernetics."
But these themes center on two main concepts: what type of person
it would take to be an inventor and what kind of inventions would be
needed to make a sustainable Iron Man. To look at these topics, we
will explore the one great evolutionary "invention" we all possess:
the human body, especially its muscles and its nervous system including the brain. We will also consider other more tangible inventions—
past, present, and future—in our quest to understand whether Iron
Man could really have been invented and, if so, what that invention
would do to the human inside.

The merger of biology, modern technology, and concepts of engineering is captured by the term "bionics." (The term also captures
fond memories of a childhood spent watching Lee Majors as Steve
Austin in *The Six Million Dollar Man*. I never understood why the
weight of a motor as it was lifted didn't rip his bionic arm right off
that human body. I still don't. But it's cool nonetheless.) The term
"cybernetics" has also been used related to this kind of research, sug-
gesting the control systems involved in combining artificial intelligence and machine-biological interfaces. The concept of "cyborg" is
also relevant here. Cyborgs show up in all kinds of pop-culture references from the Terminator of the Govinator to the Cybermen of Dr.
Who. The Terminator had a metal skeleton covered with imitation
human flesh. The Cybermen go to real extremes of biological and

machine connection and a little human biology—including an artificial nervous system—within a robotic shell of iron. Unlike our hero, the Cybermen emerged from a humanoid species on a twin planet to Earth. Those humanoids began implanting technology and artificial parts into themselves until they became full-on cyborgs and almost robots. We won't take it that far, with Iron Man, but we will take some tentative steps in that direction!

The first of many inventors we look at in this book is the man who came up with the concept of man-machine combination in the form of a cyborg. Jean de la Hire (1878–1956), an early twentieth-century French novelist, wrote a series of adventures involving a hero named Léo Saint-Clair. Léo is a man-machine hybrid whose cyborg name is Nyctalope. He possesses some artificial organs and supernatural mental powers. Nyctalope represents the first superhero written about in popular culture, preceding the great-grandfathers of comic book superheroes, such as Superman (1938), Batman (1939), and Captain America (1941). His first adventure was published in 1911 and his origin story was described in "L'Assassinat du Nyctalope"—"The Assassination of the Nyctalope"—published in 1933 and recently translated by Brian Stableford and reissued. (In true comic book style, though, Saint-Clair wasn't actually assassinated and did survive.)

Like Tony Stark, Nyctalope had an artificial heart. Or at least a heart with artificial support. But he didn't come close to Tony Stark in the complexity of his cyborg machinery. For much of this book, we will explore that machinery—the ever-changing types of armor and the men (and women) who wore them. Before we look at some representations of that armor, let's examine what major functions it performs. Iron Man's famous costume is an amalgam of an assistive device and protective armor. When we think about the kinesiology and neuroscience behind what is realistic about Iron Man, we need to be aware of both facets of the metal suit. The assistive device part is, just as it sounds, technology that assists a person in performing basic functions such as moving, lifting things, or, for the most part still fancifully, flying around. The protective function of this type of armor is much more obvious: it shields Tony Stark—and in later comics his friends and enemies—from weapons and other dangers.

In addition to these ways in which the armor assists Tony and protects him from external threats, we need to remember that the armor also provides basic support. In essence, we want to combine

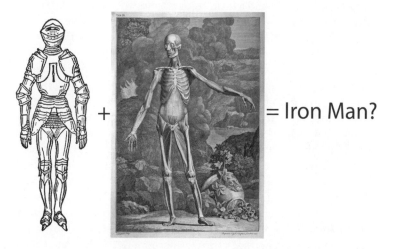

Figure 1.1. Iron Man armor seems to combine a classical view of the anatomy of the human body with the protection and support of medieval armor. *Left*, courtesy Pearson Scott Foresman; *right*, courtesy Historical Medical Books at the Claude Moore Health Sciences Library, University of Virginia.

the idea of medieval armor with the biological body (figure 1.1) shown in the painting by Dutch anatomist Bernhard Siegfried Albinus (1697–1770). Is it really as simple as shown in this figure? Does armor plus body equal Iron Man? Certainly it would produce a man in iron. But that isn't really what Iron Man is all about. The inner workings and seamless connection between the suit and the body are key.

Dozens and dozens of armors have been used since the character's debut in 1963. These include highly specialized armors like the "Hulkbuster," which was, not that the name is any giveaway, extra-beefed-up armor designed specifically to fight the Hulk. I have chosen four types of armor to highlight (figure 1.2): first, the Iron Man armor in original gray (and its gold successor); second, the classic red and gold and one special type of the red and gold, the NTU-150; and third, Extremis. The fourth is the Iron Monger armor created by Tony's business antithesis and former partner Obidiah Stane and worn by Stane and other villains. While not technically Iron Man armor, the Iron Monger is based upon the original Iron Man suit

Figure 1.2. Tony's costume has changed significantly over the years. Shown are original gray (*A*), classic red and gold (*B*), Extremis (*C*), and Iron Monger armor (*D*).

design and has been an integral part of the Iron Man mythology. It shows clearly the idea of the human inside the suit. As we will see later, Stane's armor is closest to a suit that could actually be made with current technology.

Original Gray Armor

The original gray armor made its first appearance with Iron Man's debut in Tales of Suspense #39 in 1963. On the cover of that comic (figure 1.3), there are a couple of important things to note. The first is the text that goes with the images: "Iron Man! He lives! He walks! He conquers!" Note the copious use of exclamation points! But more to the point, simply having the ability to walk is considered extraordinary. At his debut, just being able to move around in armor—so, no mention of flying at this time—was seen as something worthy of comment. The second thing to notice is Iron Man's posture. He is a bit hunched over, with his body canted forward and to the side and with his arms and legs splayed open. This is beautiful Jack Kirby art, but it is not really a stereotypically "robotic" or mechanical posture. Instead, this is an animalistic posture of something or someone crouching or otherwise preparing to move. The last thing to look at is the series of three panels shown at the left where someone (Tony Stark, as we learn in the story) has some kind of iron suit that can be taken apart and put on and off with ease. The concept of a modular suit is front and center with this panel. Remember these important things, because, as will unfold throughout the book, much of what was shown back in 1963 with Iron Man's debut in the gray armor comes closest to what is really available now, almost 50 years on.

By the way, the original gray armor didn't last long. At least not in that color. It was only around for one and a half comics. Halfway through Tales of Suspense #40, in a story entitled "Iron Man vs. Gargantus," Tony Stark turns the gray armor yellow (or golden) with some kind of metal-plating technique, and the Golden Avenger armor was born. Why the change, you may well ask? Was it strategic, or somehow related to camouflage or offensive effectiveness? No. It was used to play off of Tony Stark's persona as a flamboyant ladies' man. Just after being saved by Iron Man, one of Tony's girlfriends (Marion) asks why he wears "such a terrifying looking costume? He actually frightens people! He battles menaces like a hero in olden

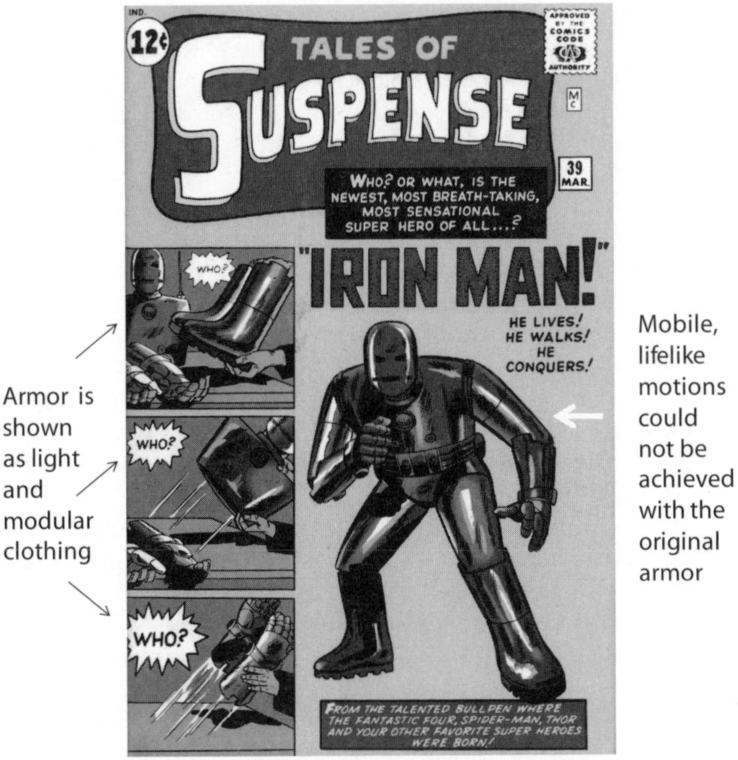
12¢
TALES OF SUSPENSE
APPROVED BY THE COMICS CODE AUTHORITY
39 MAR.
WHO? OR WHAT, IS THE NEWEST, MOST BREATH-TAKING, MOST SENSATIONAL SUPER HERO OF ALL...?
"IRON MAN!"
HE LIVES! HE WALKS! HE CONQUERS!
WHO?
WHO?
WHO?
FROM THE TALENTED BULLPEN WHERE THE FANTASTIC FOUR, SPIDER-MAN, THOR AND YOUR OTHER FAVORITE SUPER HEROES WERE BORN!
Armor is shown as light and modular clothing
Mobile, lifelike motions could not be achieved with the original armor

Figure 1.3. The first time Iron Man armor appeared in the comics was in Tales of Suspense #39 from March 1963. Notice how even at this time the armor is depicted as something that could easily be put on and taken off like clothing. Copyright Marvel Comics.

times! So, if he's a modern knight in shining armor, why doesn't he wear golden metal instead of that awful, dull gray armor?" And that was pretty much that. All that glitters is gold, I guess. (Yes, I also like Led Zeppelin.)

Classic Red and Gold Armor

The original armor was used until a December 1963 story called "The New Iron Man Battles . . . the Mysterious Mr. Doll!" (Tales of

Suspense #48, 1963), which was the debut of the "classic" red and gold armor. Since the gray armor debuted in April 1963, clearly a lot of progress happened in the evolution of Iron Man in just one calendar year. This newer armor contained many of the elements that have remained with Iron Man. At least so far. A key part of the development of the new armor was that it was more lightweight and efficient. Tony is shown contemplating "how vulnerable Iron Man is! I seem to need recharging more and more often!" and "It's this iron suit of mine! It's too heavy! Saps too much of my energy merely to support the weight!" He also commented upon how all of this had such a straining effect on his weakened heart. Faced with either giving up "the role of Iron Man—forever . . . !!" or designing "a new Iron Man costume . . . one which will be lighter in weight . . . less bulky," Tony goes with the latter. In fact one caption from this story clearly says "and so the brilliant Anthony Stark works—works—as few men have ever worked before!" His work produces the new armor.

At this stage the armor is still quite modular. The Iron Man suit can be laid out on a table (or carried by Tony in a slightly oversized briefcase, as was often portrayed) and easily put on piece by interlocking piece. An important thing to think about here, since it bears on much of what we will discuss later, is how thin and compact this armor is. Tony even points this out when he describes the torso section as "wafer-thin." Of particular significance is the headgear, which is a thin, form-fitting face shield. (We will return to the face shield in a discussion of the trials and tribulations of Tony in chapter 7.)

One thing that is never fully explained is how the motors to control movement actually work. It appears to be a much more "passive" kind of armor system but is described as a magnetized, motorized, and transistorized suit. In any event, this basic concept of Iron Man as a suit or a "costume" has persisted throughout the character's history, including the recent big-screen incarnations of Iron Man. Two images from a line of action figures sold in conjunction with the 2008 *Iron Man* movie are shown in figure 1.4. At the left Iron Man in full gear, while at the right the helmet portion is raised and the armor has been removed from the arms and chest. These images (and the figures that were used to make them) closely correspond to how Iron Man armor is depicted in the comics and movies. That is, a thin shielding that is worn very close fitting and that roughly maintains the body shape underneath. So, it is anthropomorphic armor. (We'll see later how far away from current reality this "classic" red and gold

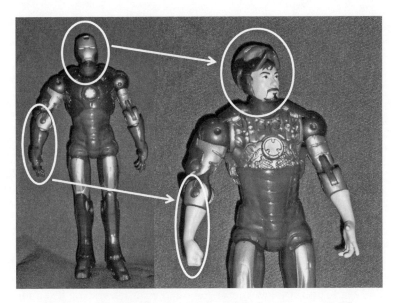

Figure 1.4. Action figure showing how thin the armor was in the Marvel Studios 2008 movie *Iron Man*. This can be seen by the circles and arrows on his figure, particularly on the face plate. It can also be seen looking at Tony Stark's bare arm. This type of armor would allow for limited cushioning of the hard impacts that Iron Man is likely to sustain.

armor is.) In contrast, the original gray armor and the Iron Monger armor are bulky and only loosely represent the shape of the human body. But they are closer to what we actually have for armored exoskeletons right now.

Telepresence Armor

The telepresence armor—actually the "Neuromimetic Telepresence Unit," or "NTU-150"— represents a unique type of suit for Tony. Many of the different armors were modified for certain uses, like doing battle against certain foes (or sometimes friends). Examples include the Thorbuster or particular armors for use in space or in the deep sea. But, the telepresence armor was a totally new type of armor technology. It no longer just protected Tony's heart and kept him alive; it also allowed him to control the iron machinery with his mind.

There are a number of versions of telepresence armor. The basic premise, though, is that it was created out of need when Tony was injured and unable to don the real Iron Man armor. Instead he devised what was essentially a remote control armor system. This was fully shown for the first time in "This Year's Model" (Iron Man #290, 1993). The background to the creation of this armor can be found in "Technical Difficulties" (Iron Man #280, 1992). In this story, Tony recounts that he built the first Iron Man suit "in order to survive a damaged heart . . . but now . . . the nerve degeneration . . . techno-organic parasite . . . is eating away at my body like an artificial cancer." The upshot is that his nervous system is degenerating. He creates a "neuro-web life support system" to help make up for the deficits in his failing body. (I just want to let you know that I attend the Society for Neuroscience annual research meeting—the largest gathering of neuroscientists in the world—pretty much every year and I have unfortunately never come across such a system. It sounds amazing.) But this invention is not enough. As we read later in Iron Man #290, this "neuro-web" works as an "artificial nervous system designed to maintain the most complex machine yet conceived by nature—a human body." Tony undergoes some fantastic procedures and when he is in the recovery room he is told "whether you'll ever recover even partial mobility—it's too early to say . . . if your system does recover, it won't be easy. You'll have to relearn even the most basic functions from scratch."

OK. So Tony is in a dire scenario. What he does next is the kind of wonderfully delicious understatement that makes comic books so awesome to read. He decides to create a brain interface that will allow him to remotely control a suit of armor. Thus is born the "telepresence unit." And, in wonderful comic book hero tradition, Tony is shown lying in bed and saying, "All I've got is the power of my intellect. Fine. That's all the power I need." (Please cue the Black Sabbath "Iron Man" power chords right here.) He gets a tech from Stark Industries to "kludge together the neural interface computer aided design and manufacture [the] teleoperator's rig" that he needs to create the suit. (How awesome is it that he had to create an interface that vastly outstrips the technology we have currently in order to build and control a robotic suit of armor that even more vastly outstrips our current technology? We will look more closely at this brain-machine

interface in chapter 3.) After Tony gets this technology squared away and gets jacked in—plugged in like a telephone into a jack in the wall—he creates the telepresence armor. Once he begins using it, Tony comments that his new tweaks have really improved the experience and that "subspace link eliminates transmission lag time. Neuromimetic system makes it feel like I'm actually there . . . That same system insures that mortal damage to the remote will result in fatal neural feedback." This newest iteration of the telepresence armor allows Tony to remotely control, in a kind of virtual environment, a fully functioning Iron Man robotic suit.

The idea of a human interacting with a virtual environment was also the central theme in some recent movies. In the 2009 *Surrogates* starring Bruce Willis, the basic theme is that remotely controlled robots are used as "surrogate people." Eventually some users die when the robotic units they control are "killed." I love one scene early on in the movie when a detective asks the police chief: "Sir, how is that even possible?" I bet you can guess the reply: "We don't know." Exactly. Anyway, the actions of the Iron Man suit were controlled by the brain activity of Tony Stark. Echoes of this story are also found in the 2009 blockbuster *Avatar*. The interface between the controllers and the Avatars is essentially a kind of biology-to-biology telepresence control and is very similar to an extreme extension of brain interface. The movie also shows the interesting biology-to-biology interface of Na'vi and the Pandoran wildlife. (The real fun part of *Avatar* is that we don't yet have the technology for the mobile ride-inside exoskeletal robots that the military uses in that film.)

When I reread the stories about the telepresence armor in Iron Man #280–290 while researching this book, I was stunned by how closely they parallel the basic workings and operational theory of current brain-machine interfaces, although most of the real ones are one-directional. That is, they send commands to control a device but cannot necessarily receive commands from that device. Iron Man's NTU-150, in contrast, was bidirectional and included information that couldn't normally be picked up by human sensory organs.

The Interior Extremes of the Extremis Armor

With a bit of foreshadowing, let's say for now that the Extremis armor comes the closest to what would be needed for the whole Iron Man concept to work with a real biological human body. By the end of this book, I hope to have convinced you of that. This concept is also the furthest away from reality of any of the armors developed so far. The Extremis armor that writer Warren Ellis created is a complete departure from everything that came before it. Not in terms of how it looks (see figure 1.2 and compare the panel C showing Extremis armor with panel B showing the classic red and gold armor), but in terms of how it interfaces with the user. We will talk in chapter 6 and elsewhere about the concept in perceptual neuroscience of "embodiment." However Extremis takes a literal approach to embodiment, becoming part of the user's body and allowing direct connection with the nervous system. The Extremis concept debuted with art by Adi Granov in the Invincible Iron Man story arc from 2005 and 2006 called simply "Extremis" and told in six parts. The origin story for Iron Man was updated for this series (Tony was now injured in the Gulf War rather than in Vietnam) and was collected in the 2007 graphic novel of the same name.

In this story, the basic idea behind Extremis lies in the work of Maya Hansen, a former girlfriend of Tony Stark and a scientist who tried to create a serum (described as a "powerful techno-organic virus-like compound" by Iron Man chronicler Andy Mangels) for a new supersoldier that became "Extremis." A terrorist group steals this serum (actually, Maya gives it to them—long story). Eventually Iron Man fights Mallen, the leader of this group, who has been amplified by Extremis. Tony suffers debilitating injuries in this exchange and has to get Maya out of prison so she can help him undergo "extremis," or a modification of his genetic expression ("rewriting the DNA"). In Tony's case, the procedure creates a kind of amplified neural network that allows him to interface completely with his armor to directly jack in and control satellites and remote computers.

The gist of this story line is that Tony and the Iron Man armor are now biologically integrated. He is in fact a cyborg. But a cyborg who can effectively turn himself on or off. With Extremis, Tony Stark has the neural interface for his armor with him at all times. It sits as a layer of electronics just under his skin and in his bones. Extremis is fascinating sci-fi on the very fringe of scientific fact.

Extremis represented a fundamental shift in how Iron Man was portrayed and really created a significant evolution for the character. Iron Man editor Tom Brevoort is quoted by Mangels as saying that, prior to Extremis, "Iron Man was a guy who had no powers and put on a suit, and when he was done, Iron Man was a guy that absolutely had some measure of powers outside of the armor, because the technology has become so integrated into him." The questions central to our work here are, How much of these concepts are realizable? And if they are realizable—however incrementally—what does it mean for the human inside (or part of) the Iron Man suit of armor? To answer these questions means understanding a bit more about the characteristics of the human body, how it works, and what it means for the human body to be interfaced with technology. Next stop, on to looking at the basic biology of Tony Stark . . . and you!

Building the Body
with Biology

WHEN THE MAN OF METAL
NEEDS TO MUSCLE IN

My transistors will operate the machine
electronically—move countless gears and
control-levers—*the iron frame must duplicate
virtually every action of the human body.*
—Tony Stark to Professor Yinsen, from "Why Must There Be
an Iron Man?" (Invincible Iron Man #47, 1972)

In motions honed to high efficiency by years of
repetition, microcircuited metal mesh armor is
slipped on, snapped into place and polarized to a
hardness that rivals titanium steel, and once more a
master inventor calls forth his greatest creation—
Iron Man!
—"Dreadnight of the Dreadnought!" (Invincible Iron Man
#129, 1979)

Iron Man in action—even just walking across a room—would turn
heads in London's Piccadilly Circus, New York City's Times Square,
or Tokyo's Shinjuku train station. However, Iron Man wouldn't be
nearly so impressive if he could only stand stock still like a statue.
Biological movement is based on the actions of muscles and, in ver-

tebrates like us humans, those muscles are layered on top of our bony skeletons but underneath our skin. For Iron Man, the link between muscle activity and motorized actions of his mechanical exoskeleton has to be almost symbiotic.

Normally, when Anthony Edward Stark wants to make a movement, a chain of commands begins in his brain and finishes with the contraction of his muscles and the actions of his body. For example, think back to that excited reach you made to pull this book off the shelf and then carry it, full of hope, to the checkout line. (Or, for those of you so inclined, the motions you made to click on a computer to order it online. That's cool, too.) When you decide to make a movement like lifting up your left hand, a series of commands are relayed from neurons in your brain, down to those in your spinal cord, and then out from the spinal cord to the muscles themselves to make them contract. That chain of command is an inherent part of Tony Stark's and your nervous systems.

Muscles sit quietly inside our bodies, leaping into action only when we need them to do something. And, honestly, we need them to do something pretty much all the time. We humans have an awful lot of muscle in our bodies and they make up a whopping 40% of our total body weight. So, Tony Stark, whom Marvel Comics lists at about 6'1" in height and about 100 kilograms (225 pounds), is packing about 40 kilograms (90 pounds) of muscle! You—and Tony—have three kinds of muscle in your body: smooth (like that found in your gut), cardiac (found in your heart only), and skeletal (found in the muscles that move your skeleton).

Skeletal muscle is probably the type you think of right away when someone shouts "muscle." Your body has 639 skeletal muscles ready and willing to act during deliberate voluntary actions, during automatic activities like walking, and during reflex corrections to movements. The focus in this chapter is on how Tony Stark would use these muscles to help him control a fancy robotic suit of armor and on the chemical reactions that make these movements possible.

First, let's look at some of the major muscles that are important for moving our arms and legs and giving us stability. In figure 2.1, panel A shows muscles on the front of the body and panel B shows muscles in the back. All 600-plus of these muscles will need to be protected and enhanced with the Iron Man suit. Of great importance is figuring out how realistic it might be for Tony Stark's nervous system to be linked to the muscles that move his body and to the suit that surrounds it.

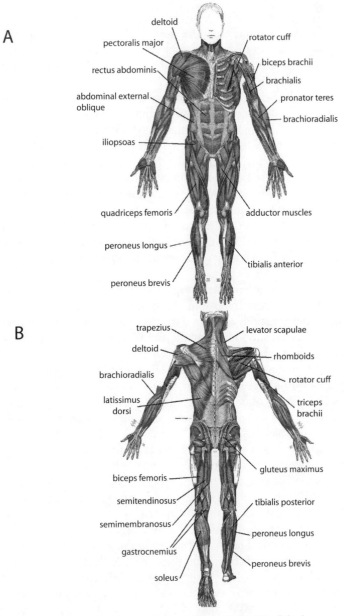

Figure 2.1. Some muscles on the front (*A*) and back (*B*) of the human body. Images from *Gray's Anatomy* modified by Mikael Häggström.

In such a suit the connection between brain and armor would need to be so good that the brain could actually control an instrumented robotic suit of armor, as if that suit of armor were a human body. In effect, Tony would have to create an anthropomorphic suit. That's the same "anthro-" as in anthropology and means relating to things human. So, it must be a suit that is meant to look and act like a human. Such a degree of connection wasn't needed in the first versions of Iron Man armor or even in the Iron Monger armor. For the level of interface between man and machine shown in recent comics, though, that type of suit would be needed.

To set up our exploration of this interface, we first look at how your own body is controlled. How does the nervous system work to produce movements? What kinds of signals are used and how does it all work together? The "motors" of your body are your muscles. In neuroscience the term "motor control" is used to describe how the brain controls movement. But real motors aren't usually part of movement. Could muscle control be used to mimic the control of motors in a real machine? To answer these questions, let's review the basics of biological activity and of the anatomy and physiology of the human body. This review will help us when we look at the topics of controlling robots, robotic exoskeletons, and similar inventions.

Muscles Alive! They Twitch! They Contract! They . . . um . . . Make Heat?

Let's compare the overall structure and function of the limbs in our bodies to those of a robotic exoskeleton. The basic biological principle is that the bones of our skeletons support our bodies, the muscles move our bones, and our brains command our muscles. The sum of all the muscle activity and bone movement is the movement of our whole bodies. In the case of a robotic suit of armor, the new skeleton is on the outside and the "muscles," in the form of motors or actuators, are on the outside of the new skeleton. So, we are dealing with two sets of supporting skeletons and two sets of muscles. For this whole enterprise to work, there has to be a link between these two sets. How that could possibly occur can best be understood by thinking through how the human nervous system functions. The cells of your nerves and muscles are what is known as "excitable tissue."

This refers to electrical excitability. Storing electrical activity of your cells is similar to the way a battery functions, except the "discharge" of your neurons powers information transfer instead of a flashlight or a Nintendo DS game. Which is pretty cool, in my estimation.

Nervous about Neurons

Neurons are excitable cells that generate and carry electrical signals. (Don't worry. You will get a look at a neuron from the motor system in figure 2.3 later in the chapter.) The electrical signals result from an unbalanced concentration of ions on either side of the nerve cell membrane. Quite a lot is accomplished with just three household-sounding ions: potassium ($K+$), sodium ($Na+$), and chloride ($Cl-$). By the way, bananas, sweet potatoes, and halibut are all excellent sources of dietary potassium intake. You probably get most of your sodium and chloride in the form of table salt, $NaCl$. Other trace sources of chloride are olives, tomatoes, and celery, and sodium can be found in barley and beets. In addition to these three main players, you also have some special other ions inside your cells. Actually, these ions are found inside and outside of your neurons, but a dominant concentration is maintained in your cells by cellular pumps.

The cellular pump (or exchanger) is usually referred to by its biochemical name, Na-K ATPase. This cellular pump works in a similar way to a revolving door that helps move people inside and outside of a building. Put the potassium ions in the crowd moving in and sodium in the crowd moving out and you get the basic picture, with one twist. The Na-K ATPase revolving door doesn't create equal opportunity openings. For every two $K+$ ions that are pumped in three $Na+$ are pumped out. As a result, at rest, sodium and chloride ions are much more highly concentrated outside your cells, while potassium is more highly concentrated inside. Because ions have either a positive or negative charge and they aren't evenly distributed across the cell membrane, you wind up with an electrical difference between the inside and outside of the neuron. This is called the "membrane potential."

A way to appreciate this potential difference is to think about a physical example. Imagine a tub of water. Now take a glass and put it into the water upside down so there is air trapped inside it. As you push it down into the tub, the glass will have some water enter it as the water pressure overcomes the air pressure in the glass. This is similar to the way that the membrane potential increases and decreases across your neurons. This changing level signals information flow in the nervous system. All cells have resting membrane potential differences, but it is only excitable cells in nerve and muscle that can generate changes in membrane potential and transmit those changes as information.

Muscles as Motors—How Do Tony Stark's Muscles Work?

The spinal cord is highly organized. Sensory innervation of—or bringing nerves to—the skin on different parts of the body corresponds with different levels of the spinal cord. The relationship between nerves in the spine and what are called "dermatomes" is shown in figure 2.2. The organization of the dermatomes is typically thought of using the four regions of the spine and spinal cord. From head to "tail" these are cervical, thoracic, lumbar, and sacral. This concept of dermatomes is useful for diagnosis in clinical neurology. A good (bad?) example is with back injuries. If you or someone you know has had a disc herniation, they likely had some changes in how sensory information from the skin on the legs was relayed. For example, about 20 years ago I had three herniated lumbar discs at L3-L4, L4-L5, and L5-S1. As a nice reminder, I now have patchy sensation on the skin of my lower legs associated with the dermatomes for those spinal levels.

Activity in the nervous system leads to the activation of muscle. If Tony Stark decides he wants to pick up a laser-guided tool of some kind—maybe an arc welder—to make a modification to the Iron Man armor, a command will arrive at the motor cortex of the brain. This is the main movement-control area of his brain and the place where the commands that are sent down the spinal cord to trigger muscle contraction come from.

There is also the issue of motor innervations to "motoneurons"— short for motor neurons—for different muscles in the body. The motoneurons are also organized at anatomical levels. For example, when the signal descends to the part of the spinal cord where the nerve cells for the muscles involved in reaching are found, the cervical region in this case, activation of those cells commands the muscles to become active. For biological systems like your body, the lowest level

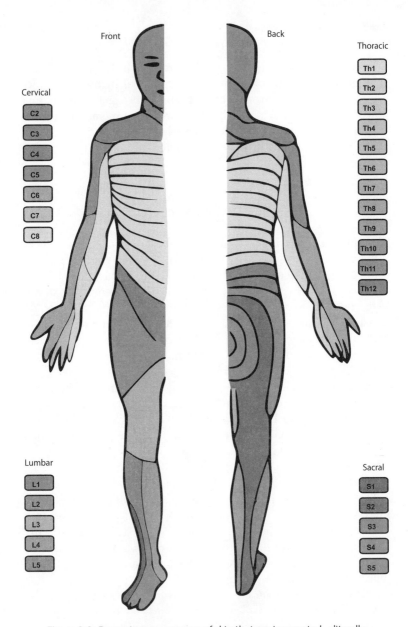

Front   Back

Cervical

C2
C3
C4
C5
C6
C7
C8

Thoracic

Th1
Th2
Th3
Th4
Th5
Th6
Th7
Th8
Th9
Th10
Th11
Th12

Lumbar

L1
L2
L3
L4
L5

Sacral

S1
S2
S3
S4
S5

Figure 2.2. Dermatomes are areas of skin that are innervated—literally, supplied with nerves—by the sensory fibers from nerve roots in the spinal cord. Note that each dermatome is named according to the spinal nerve supplying it. Although there are seven cervical vertebrae, there are eight cervical dermatomes. Courtesy Ralf Stephan.

of control for generating force is using what is called a "motor unit." This unit is the nerve cell in the spinal cord and all the muscle cells (or fibers) that it connects with. So, when a muscle contraction occurs, it results from activity in many muscle fibers. The muscle fibers have proteins within them that regulate contraction and also produce the actual contraction forces. In figure 2.3, there is an example of a motor unit showing just a few muscle fibers. The motoneuron is at the top of the figure showing the cell body and dendritic tree (the filament-like bits at the top) in the spinal cord with the axon shown

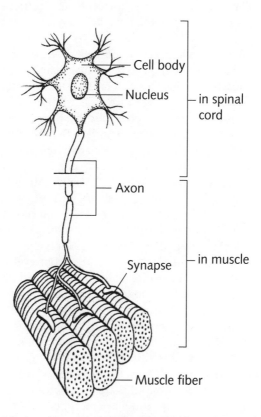

Figure 2.3. A motor unit—the basic functional unit for movement—consists of neurons (cell bodies) in the spinal cord, along with the extension of the axon out to the fibers in the muscle. Courtesy Johannes Noth (1992).

extending out to the muscle. The things that look like hot dog buns on the outside of the axon are the sheaths of fatty insulation called "myelin" that help keep signals moving quickly.

Once the electrical signal from a motoneuron arrives at the synapse (called the neuromuscular junction), that signal becomes chemical. The term "synapse" was first used by Sir Charles Sherrington, a Nobel Prize–winning physiologist (and, as readers of *Becoming Batman* already know, one of my superheroes of science). At the synapse for the neuromuscular junction, the chemical neurotransmitter acetylcholine is released, crosses the gap, and leads to depolarization of the muscle fibers. This turns the signal into an electrical one again and causes the release of calcium ions that serve as triggers for muscle contraction.

Muscle fibers are composed of different proteins. Some of these proteins have a direct role in producing muscle force and are called "contractile proteins." Others have an indirect role in regulating contraction. The contractile proteins are actin and myosin molecules. Motor units come in "sizes," which means that the number of muscle fibers controlled by each motoneuron differs. But the number is variable in a logical way and can range from about ten muscle fibers for one motoneuron up to thousands of muscle fibers. Keep in mind that when a motoneuron is commanded to be active, all the innervated muscle fibers must also be active. Your motor units (or at least the muscle fibers innervated by them) are real team players! This means that a small motor unit with ten fibers will produce less force than a large unit with a hundred fibers. Not surprisingly then, we find that the motor units controlling the movement of your eye (which has a very small mass) are the smaller units and those controlling your much larger (and heavier) leg muscles are the larger units.

If Tony needs to push more forcefully with a handheld laser— that is, he wants to change force production in his muscles—his motor commands will make more motor units become active (called "recruitment") or the motor units that are already going will be active at a higher frequency. Both of these things happen more or less at the same time, except motor units are recruited and then activated at a higher rate. Then more units are recruited and made to discharge higher and so on. Meanwhile, the force of contraction will be steadily increased. This is something you wouldn't notice at all, but it is a great example of matching between the output of the nervous system and the mechanical ability of your muscles to produce force.

Tony Stark's motor units come in two basic types depending on how fast they contract (or "twitch") and how fatigable they are (how long they can keep contracting before having to stop). These two main types are called, rather unimaginatively, type I and type II or "slow twitch" and "fast twitch." The fastest twitch and most forceful units are the type II. The slightly less strong and slower twitch units are the type I. When Tony is manipulating that laser, he would be bringing into activity more and more units at different frequencies. If he had to hold it for a prolonged time, he would begin to experience fatigue. His muscles would start to ache from the pain detected by the metabolic processes occurring and his arm might begin to shake or appear to vibrate a bit. This is called "physiological tremor" and wouldn't be particularly helpful if accuracy were needed. So, he would have to rest a bit between his efforts.

Your muscles are a really efficient kind of biological motor. If we were to compare them with real technological motors, it would probably be most useful to compare power output based on weight. According to Steven Vogel and his work on human muscle, your skeletal muscle produces about 200 watts per kilogram (90 watts per pound). The steam pump, debuting during the industrial revolution in 1712 in the hands of Thomas Newcomen and refined by James Watt in 1775, produced just over half of that at about 50 watts. That doesn't quite match 200 or 500 watts per pound for a car or motorcycle engine, but it still is pretty good. By the way, a jet aircraft turbine clocks in at approximately 2,500 watts per pound. All in all, for a squishy bit of biological material, our muscles do pretty well.

Probably the main thing you think about concerning your muscles is how much force they produce. You might be surprised to learn that there are a number of oddities that occur during muscle activation and force production that introduce a few wrinkles into what we are able to do and how we (unconsciously) do it. Imagine a forklift tractor with the loader on the front. The hydraulic piston that helps raise and lower the load on the tractor behaves in the same way each time it is used. This is a highly linear system, and it is tempting to think of muscles as working in a similar way. But, muscles operate in a nonlinear system and have some peculiarities. The force that your muscles produce depends on the length of the muscle fibers and the speed at which the muscle is contracting. The details of the relationship can be a bit complex but, generally, the faster a contraction is occurring, the less force that can be produced. During a slower contraction, more force can be produced. This is called the "force-velocity relation." You can flip this around into a load-velocity relation and think instead about how fast you could contract arm muscles to move your arm when holding a light weight versus holding a heavy weight. The lower the weight, the faster you can move, since, as we've noted, less force is needed to move a lighter weight. All of this applies to what are called "shortening contractions," which are when muscles are active and are shortening (also called "concentric actions").

If you are sitting down right now, you could do a shortening contraction with your knee extensor muscles (your "quads," or quadriceps, on the front of your upper leg) by raising your leg until your foot is parallel with your hip. If you hold it out at complete knee extension, you are performing a constant length, or "isometric," contraction. When you slowly bring your foot back down and let your knee extensors relax, you are performing a lengthening, or "eccentric," contraction. A lengthening contraction produces the highest forces, and this force goes up the faster you move.

During everyday tasks, your muscles are constantly doing both shortening and lengthening contractions. For example, when you go up stairs, the knee extensor muscles are doing a lot of shortening actions. When you come down, they perform lengthening contractions. But your muscles are about 30% more efficient going downstairs than they are going upstairs. The actual force that moves those bones comes from the contractile parts of the muscle fibers as well as the connective tissue that holds the fibers and the muscle as a whole and keeps everything connected to the tendons. All of those pieces together have some elasticity, and it is because muscles use this elasticity when they contract while stretched that makes them more efficient. Elasticity in muscle and tendon has important implications for moving Tony Stark and his Iron Man suit around.

How Can Iron Man's Motors Mimic Tony's Muscles?

A main theme of this book is examining the amplification of human performance by the use of technology. Generally when we think of robotics and control issues inherent in the Iron Man suit of armor, this means thinking about the state of the technology in powered devices. Let's consider another state-of-the-art example that has to do with something more day to day: sports biomechanics and performance using modified prosthetics. The term "prosthesis" simply refers to an artificial extension of a part of the body, usually a part that is missing or has been lost due to disease or injury. Unfortunately, and rather grimly, there is a strong link between research into amplifying the performance of amputees using prosthetics and assistance for soldiers. Soldiers in battle are and have always been at risk for losing limbs during combat. There has historically been a big technological advance in prosthetics development following large military operations in many cultures. For example, there was a huge upswing in the need and interest in prosthetics during the U.S. Civil War. In recent times, this has been greatly exacerbated by the role that "improvised explosive devices," or IEDs, have played in the wars in Iraq and Afghanistan. Just like the horrible toll that landmines have and continue to take, these devices lead to many limb amputations.

As seems to be the case with so many things, the concept of prosthetic limbs can also be traced back to Egypt, where an artificial toe was created and used around 1500 BC. The first known and effective prosthetic hand that is relevant here is that of a German knight named Götz von Berlichingen about AD 1500. His hand was injured by cannon fire during battle and he had a mechanical one created. His use of the artificial hand gave him the nickname of the "Iron Hand" and was written about in the Goethe play of the same name. I cannot explain how amazing it was to discover a knight called "Iron Hand" when researching for this book about Iron Man. It could not have been better. As you can see in figure 2.4, the hand actually had a mechanism that allowed the fingers to flex, so it was really quite advanced for its time. If we zip forward to 1800 and the Napoleonic wars, we come across an above the knee prosthetic leg called the "Anglesea leg." It was used by Lord Uxbridge, who was a cavalry officer for the duke of Wellington (who, by the way, was also known as the "Iron Duke"!). This leg had an advanced mechanical design that allowed the foot to automatically rotate upward while the knee bent during the swing phase (when the foot is off the ground) of walking. This is just like the way the foot moves during walking across the ground normally, and this movement allowed for clearance between the foot and the ground to prevent tripping. This basic design persists into modern day in many prosthetic limbs.

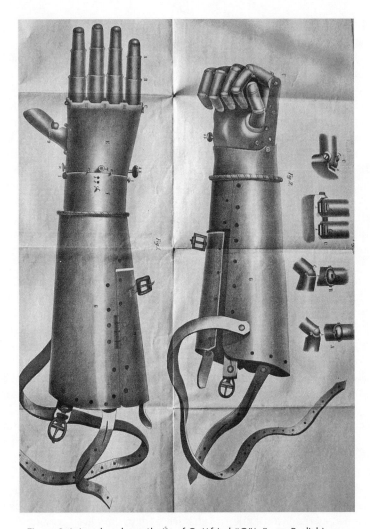

Figure 2.4. Iron hand prosthetic of Gottfried "Götz" von Berlichingen (1480–1562). This mechanical prosthetic hand began use in 1504 when von Berlichingen lost his arm in military action. It could be used for a range of activities from wielding a sword to holding a pen.

Does Iron Man Have a "Spring" in His Step?

Let's now also consider a sports example, namely, the use of specialized prosthetic legs by world-class sprinter and double-amputee Oscar Pistorius. There are two major bones in the lower leg called the tibia (your shin bone) and the fibula (the one that runs along the side of your leg—you can feel the bottom of it as the outside of your ankle). Both bones are needed to provide stability for the foot and to form the ankle joint. Oscar was born without fibulas in both legs, and there was really no chance that he would have been able to stand and walk. So, when he was still an infant (about 11 months of age), his legs were surgically amputated. This may sound dramatic and clearly caused a permanent loss. Removing the lower part of his legs allowed him to use prosthetics that could carry his weight and could be used for walking and, essential for what he does now, running. Very fast. Oscar is now the most dominant double-amputee sprinter and has won several gold medals at the 2008 Paralympics in the 100, 200, and 400 meter distances. His prosthetic legs are termed "blades" (you can see why in the image of him running shown in figure 2.5), but the technical name of these carbon-fiber prosthetics is the "Ossur Flex-Foot Cheetah." The "flex-foot" part of the name is key to how the legs help with performance. Another component that helps Oscar's performance is that the blades are so lightweight. In fact, the prosthetics work so well that, when combined with the superb sprinting physiology and mechanics of Oscar Pistorius, they created a controversy in track and field in 2007 when he first competed against runners with intact limbs. He did so well that claims were made that his "blades" gave him an unfair performance advantage over runners who had to use their own legs!

As a result of this, the International Amateur Athletic Association (or IAAF) as the governing body for track and field banned the use of performance-enhancing technology, included the use of the "blades" and ended any idea of Oscar's competing in the 2008 Olympics. In fact, IAAF Rule #144.2 prohibits using "any technical device incorporating springs . . . that provides the user with an advantage over another athlete not using such a device." As posed by Brendan Burkett, a professor of biomechanics at the University of the Sunshine Coast in Australia, "Does the technology create an unfair advantage for the Paralympian when competing against able-bodied Olympic athletes?" Burkett raised a number of very interesting

A

B

Figure 2.5. South African Paralympic runner Oscar Pistorius using his lower leg lightweight carbon-fiber "flex-foot" prosthetics (A) while running in Iceland; running stride using "flex-foot" blades (B). Panel A courtesy Elvar Palsson; panel B adapted from Weyand et al. (2009a and 2009b).

"thought problems" when trying to work out the issue of technology in worldwide sports competitions like the Olympic and Paralympic games. If technology can play such an important role in human performance, what about the problem of equal access to all competitors? He points out the issue of access by highlighting the stunning fact that the winner of the 1960 Olympic Marathon was an Ethiopian named Abebe Bikila—who ran barefoot. With access to technological advances like running shoes, Bikila's time would have certainly been even better than the winning margin he did produce. There are numerous more recent examples, such as the use of specialized fibers in tracksuits or swimming suits that can change the drag and friction experienced while running and swimming. When we come back to the example of the "blades" of Oscar Pistorius and the issue of using prosthetic limbs to enhance performance, for a minute it does seem a bit bizarre to question the performance of a person who has something that should limit running performance—that is, no lower legs. However, if we think a bit further maybe there is something to this. Can technology be used to not just restore or replace function that is normally part of the body, like having legs to walk and run with, but to actually enhance performance beyond the level of those with intact bodies?

To fully appreciate why this design can be so useful, let's consider the way walking works. Walking and running are like falling forward and continually catching up to our falling bodies by taking our next steps with our legs. This is often visualized by the movement of the body's center of mass (COM) across the ground. On average, your COM is approximately located in the middle of your abdomen behind your belly button. The movement of the body is often described as an inverted pendulum, because, once the next step is taken, the COM kind of vaults over the place on the ground. However, a key part is that there is a springiness to this vaulting. As your weight moves onto your leg, there is a compression due to the motions at the hip, ankle, but mainly the knee. There is a large elastic—or springy again—property to your muscles and tendons, which is what is tapped into during walking (and other behaviors). Because of this, a biomechanical model of the legs during walking often shows a spring where the knee is. Then the limb is described as having a certain stiffness or compliance. By changing muscle activity (or more generally by wearing different shoes), you can change this stiffness in a step-by-step way. A portion of this is carried out by your reflexes, particularly the stretch reflex (figure 2.6). Quite a lot of our efficient walking and hopping is due to the way our nervous system assists our movement using our stretch reflexes. Stretch reflexes occur when the lengthening (or stretching) of a muscle activates sensors in the muscle itself. Those receptors send signals back to the spinal cord, which then reinforces and supports the activity of that muscle.

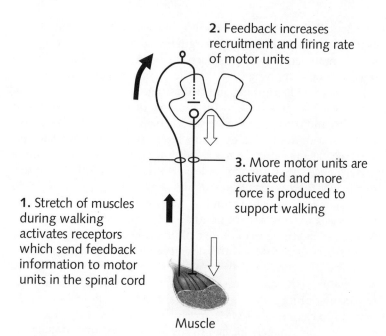

2. Feedback increases recruitment and firing rate of motor units

3. More motor units are activated and more force is produced to support walking

1. Stretch of muscles during walking activates receptors which send feedback information to motor units in the spinal cord

Muscle

Figure 2.6. Simple reflex pathway ("stretch reflex") to support contraction of the lower leg muscles during running and walking. Courtesy David Collins.

A dynamic change occurs in your walking or running when you go from running barefoot on a soft sand beach to running on a hard pavement foot path. This can trip you up—both literally and figuratively—sometimes. Once I was running across an airport trying to make a connection due to a delayed flight. I was running on a very soft springy moving walkway and just kept right on running as I came off the other end and landed on the hard tiled surface of the normal airport walkway. The jarring that I experienced felt like it had loosened all my teeth. It only took one step and then my body—set by my brain—had adjusted. You may have experienced this kind of thing by actually running in the sand. You get a similar kind of mismatch if you are running or walking down a staircase—typically at night—and then misjudge where the last step is. That can be very jarring too, and it is caused by that mismatch in what is expected to occur and what actually happens.

Dan Ferris and his colleagues have been researching aspects of this very problem—the dynamic changes in the control that the body can produce—for quite some time by using exoskeletons fitting over the legs to modify natural movement and to examine how neural control adapts when this happens. Their work has shown that when we run or hop on different surfaces, the stiffness of our legs is adjusted to compensate for the surface. This means there is less disturbance to the motion of the whole body. Think about running on a sandy beach compared to running on a paved road. Or running as a human and running as Iron Man. The general idea is kind of like having an adaptable shock absorber that is tuned exactly to what kind of shock you will experience. Figure 2.7 shows the lower limb working like a "shock" absorbing spring during walking and running. Panel A demonstrates how movement of the ankle and knee (and activation of muscles acting at those joints) produces a spring-like effect that the body works on while walking and running. As you walk or run, remember that your COM is always rising and falling slightly. This is controlled by the "stiffness" of your legs and shown by the circle on the top of the spring in the figure. The stick diagrams (panel B) show the movements at the ankle, knee, and hip. The exoskeleton (panel C) is used to change the mechanical responses of the ankle joint. Ferris's research has been tremendously successful in advancing our understanding of how the body works during walking, running, and hopping.

If you can appreciate the mechanics of walking in this way, it becomes pretty clear why there has been so much controversy over a device that changes the properties of the legs. Essentially the "blade runner" prosthetic used by Oscar Pistorius dramatically increases that springiness. It stands to reason that this would also make performance much better. However it was mostly an intellectual argument until recently.

Because of all the legal issues that arose from the track and field controversy, it was necessary to see if there was any science behind the contentions. Several movement scientists were called in to measure the physiological cost of walking and running and the biomechanics of using prosthetics. Peter Weyand and colleagues performed studies comparing the speeds, metabolic energy cost, and biomechanical characteristics of the double amputee running with the blades with those of elite, high performance track athletes. Incredibly, the use of the blades allowed for almost equivalent performance to

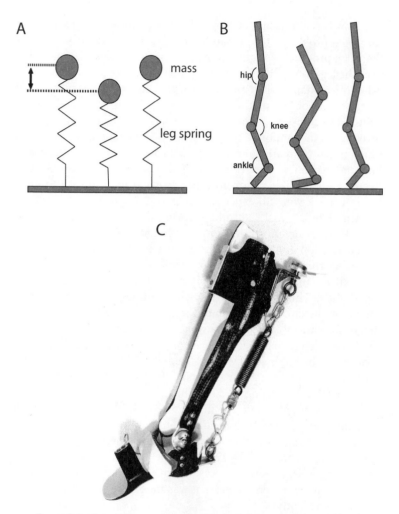

Figure 2.7. The circle represents the body center of mass suspended over the legs, which work as springs while walking (A); the "leg spring" comes from controlling muscles that cross the hip, knee, and ankle (B); exoskeleton worn over the leg used to change the mechanical responses of the ankle joint (C). Courtesy Daniel P. Ferris.

elite level runners. This is stunning in terms of how technological advance in prosthetics can give rise to such a leveling of the playing field. However, it has raised some additional concerns about whether the use of such technology may actually allow for increased performance, that is, for performance better than that of people with intact limbs.

The bottom line of all the scientific analysis is that the carbon fiber "blades" could significantly enhance performance. This is largely

Figure 2.8. Parade performer in Disneyland, California, using a lower limb exoskeleton to amplify the springlike activity that the legs produce normally.

because the blades allow for a far more efficient running pattern. The blades are actually much lighter than the lower legs they replace, which means that the legs can be moved about 15% faster than the highest performance of sprinters with intact legs—including 2008 double gold medal winner Usain Bolt of Jamaica. Also, the same overground running speeds could be obtained using the blades while applying about 20% less force into the ground. Overall the "springiness" of the blades meant that only about one-half of the muscle force needed for sprinting at the same speeds with intact limbs was needed with the prosthetic legs.

If Oscar were a marathon runner, there would be different issues. On New Year's Day in 2010 American ultradistance runner Amy Palmiero-Winters won the "run to the future." She covered 130.4 miles in 24 hours, making her the first person with a prosthetic lower leg to qualify for the U.S. track team. Her situation is different from Pistorius, because the mechanical benefits Amy might get from a single prosthetic leg don't really help with long-distance running.

If you keep your eyes open, you can see small-scale applications of assisting human movement with technology in many different places. I took the picture of a performer during a family trip to Disneyland in 2009 (figure 2.8). The basic ideas we have been discussing are clearly shown in how he used spring boot pogo sticks to amplify movement. The main point of this as it relates to Iron Man is that even simple devices can augment and improve function in people with amputation—and those like the Disneyland performer who are just trying to have fun.

What is important to consider is that really efficient machine-based locomotion should probably mimic what we do when we just walk around. So, an Iron Man armored suit should do the same. After all, as Tony Stark said while testifying at the "Weaponized Suit Defense Program Hearings" (shown in the *Iron Man 2* movie), his device is really just "a high-tech prosthetic." Now let's look at some other prosthetics that link directly to the nervous system.

Accessing the Brain of the Armored Avenger

CAN WE CONNECT THE CRANIUM TO A COMPUTER?

Undergoing the Extremis Procedure remade my body from the inside out. Long story short, my body was turned into a kind of computer designed to interface with the Iron Man. There was no longer a division between me and the suit. My brain . . . evolved, I guess. Into a kind of hard drive.
—Tony Stark, "World's Most Wanted, Part 2: Godspeed" (Invincible Iron Man #9, 2009)

The original version of the Iron Man armor was designed to preserve my damaged heart. The obvious next step was to extend the suit's preservative capabilities to an even more critical organ. . . . Not so much my brain per se, as my cognitive neurofunctions and basal personality structure.
—Tony Stark speaking out loud about new changes to the armor, from "Hypervelocity #2" (Iron Man, 2007)

"Coffee pot on!" Imagine if you could, upon awakening, simply have that thought and your coffee maker would go on in your kitchen.

Or your kettle, if you prefer tea. Imagine if your thoughts could be transformed into the actions of machines. In the last chapter, we looked at muscles and how they make us move after receiving electrical commands from the nervous system. In this chapter, we look at how we can tap into the command signal from the nervous system to not just make muscles contract but to trigger powerful motors to move robotic suits of armor. Another focus of this chapter is on what happens when that chain of command is broken, which relates to that imaginary ability to turn on your coffee maker with thoughts alone. We will discuss how that works by giving some examples of prosthetic arms and legs being powered with the mind.

These real-life bionic men and women bring us to the fascinating field of neuroprosthetics. This term refers to devices that are used by or implanted into a person to improve sensory, and, in some cases, cognitive abilities, including retinal and cochlear implants to augment sight and hearing, spinal implants to relieve pain among other things, and implants to assist with bladder control. Something that Tony Stark would be able to use would be the implants that are being developed to control movement of an object by simply thinking about it. These inventions are in their infancy but are expected to help paraplegics and quadriplegics and others with severely limited movement. And who knows what else they might soon be able to do?

Some of the possibilities of the future are being revealed by pioneers such as Kevin Warwick, a professor of Cybernetics at University of Reading in the United Kingdom. Warwick is known for many things including "Project Cyborg"—his attempt to have implants placed into his body that can be used to control other devices. One of his early efforts was to have a computer chip implanted that could be detected by sensors outside his body to then turn on lights or other appliances. Later, he had an electrode array implanted in his arm. This array took information from a nerve in his forearm, which was used to control a robot arm "directly" and eventually over Internet relays. Warwick summarizes his approach in his book *I, Cyborg*. He has had some exciting successes with this approach, but the problems that have arisen in doing more complex tasks for longer periods highlights how difficult interfacing humans and machines actually is.

Remember from chapter 2 that when you want to do something, you must activate your muscles. When you integrate your body with a robotic machine, you must skip the muscles and go directly from the output in the spinal cord straight on to the machine. In a way, we are doing some wire-tapping in the nervous system. The first wire tap we want to set is from the spinal nerves. Later, we will also talk more about even tapping into commands from the brain itself using a special kind of neuroprosthetic—a brain-machine interface.

The general principles for neuroprosthetics, as well as other types of prosthetics, are similar for both amplifying human performance and for replacing it. Many advances in prosthetics have improved the "usability" and the look of the devices. As we saw last chapter, however, the limbs themselves don't actually function in the same way as in an intact situation. An idea that arose early on in the field of neuroprosthetics was controlling a motorized limb using commands from the nervous system. In other words, tapping into the commands that would normally activate the muscles themselves. Instead of needing to figure out exactly what the complex sets of commands should be for a given movement, a simple and elegant approach is to instead just use the input itself.

Most neuroprosthetics detect signals from the person's nervous system and relay these signals to an electrical controller inside the device. The biological signals could be electrical activity of muscle detected from electrodes on the skin or implanted in muscle, nerve signals detected by implanted electrodes, or even electrode arrays in the nervous system that have the nerve cells growing through them. In this way the controller of the neuroprosthetic is literally connected to the neuromuscular system and to the device. The commands from the person can then be detected and relayed to the device to make it do whatever it is supposed to do to replace the lost function.

Monitoring Muscle and Nerve to Make Motors Move

If damage to the nervous system, such as an injury to the spinal cord, occurs, electrical stimulation can make the muscles contract even when the nervous system itself cannot provide the command for the contraction. Since the point of the stimulation is to help with functional movement, a term that arose is "functional electrical stimulation," or FES. FES is now more broadly used to refer to using electrical currents to produce or suppress activity in the nervous system. When similar stimulation concepts are used to more generally alter activity within the nervous system, it is often called "therapeutic electrical stimulation."

An example of a really useful FES device is the "WalkAide." I bet you would never guess it helps with walking. The WalkAide is a small battery-powered device used to stimulate a nerve that activates the muscles that help flex your ankle to bring the top of your foot up when you walk. If your nervous system is working well, you probably pay this no attention at all, but the clearance of your foot over the ground when you walk is a hugely important issue during walking.

Because it is energy inefficient to pick the foot way up off the ground while walking, your nervous system activates your leg muscles so that the bottom of your foot just clears the ground by less than an inch. This is all fine and good until something like a spinal cord injury or a stroke occurs. These disorders lead to weakness of the muscles, particularly those that flex the ankle, and suddenly walking is much more difficult. You may have experienced a brief example of the outcome of this phenomenon, called "foot drop" or "drop foot." Sometimes a piece of sidewalk will be broken and jutting up higher than the clearance of your foot while walking. You don't see it as a major object visually but then you scuff your foot and may trip. Well, after a stroke this is common even when the place a person walks on is level and smooth. Two simple things can be done: one is to swing the leg out and around when walking (so the affected leg arcs to the outside and forward—called "hip hiking"). The other is to wear an external brace of metal, plastic, or graphite that holds the foot at about a right angle so it cannot scrape against the ground. Neither is really a great approach. So where does FES fit in?

With the WalkAide (figure 3.1), based on the acceleration and the angle of the leg, a sensor detects the correct time during walking when flexing the ankle should occur but doesn't because the nervous commands are lacking due to injury. So, the WalkAide applies electrical stimulation to the nerve, activating the muscles that flex the ankle and allowing the person to pick up the foot. If you reach down the outside of your knee, you will come to a little bump on the outside of your lower leg just below the knee. This little bump is actually the "head" of the fibula—the leg bones we talked about earlier with Oscar Pistorius. Right near the head of the fibula is the common peroneal nerve that innervates the major flexor of the ankle, the tibialis anterior muscle (see figure 2.1). The WalkAide appropriately stimulates (and just as importantly) stops stimulating this nerve in order to get the foot moving better during walking. It is small, easy to use, and requires 1.5 volt AA batteries. The stimulator unit is shown

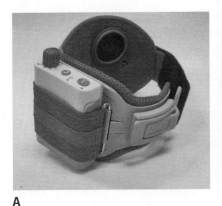

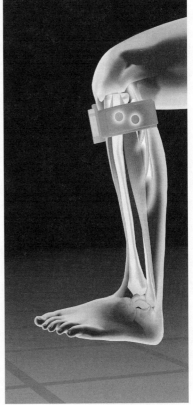

A

B

Figure 3.1. WalkAide neuroprosthetic stimulator, which corrects a condition known as drop foot. The stimulator (*A*) activates the common peroneal nerve, which causes activity in the muscles that lift the foot (*B*). Images courtesy Innovative Neurotronics.

in panel A and a drawing of how it would help move the ankle by stimulating the ankle flexor muscles is shown in panel B. The bulk of the initial WalkAide technology was developed in the laboratory of Richard Stein at the University of Alberta in Edmonton, Canada, during the late 1980s and early 1990s. I was doing my doctoral training in neuroscience there in the 1990s and got to see some of the early prototypes from this work up close. It was captivating then and still is now. But back then it was an idea in development. Now it is an actual product that people can get and use to help improve their walking.

The WalkAide is an example, then, of a neuroprosthetic. It represents a medical device that helps improve (or replace in some cases) bodily function that has been lost due to accident or disease. Other kinds of neuroprosthetics include stimulators for bladder and bowel control, deep brain stimulation (we will come back to this later on in this book—so stay tuned), and cochlear prosthetics. Usually neuroprosthetics require insertion and implantation of electrodes into the body near the nerve or muscles that are targeted. But, the WalkAide is an example of a neuroprosthetic that doesn't need any implantation.

Cochlear prosthetics to improve hearing are the most commonly applied and utilized neuroprosthetics. They represent an instructive example of how continuing evolution in the fields of biomedical engineering and neuroscience from the 1950s to now have dramatically improved neuroprosthetic devices. Originally cochlear implants were very large and had external pieces fixed to the body that were wired to parts implanted into the inner ear. Now they are small directly implanted devices. This really nicely parallels the concept of the cardiac pacemaker and defibrillators. We will discuss this in chapter 7 when I put a new twist on Iron Man's origin story!

Another interesting example of this kind of FES is a neuroprosthetic for improving hand function developed by Arthur Prochazka (coincidentally also at the University of Alberta). This "bionic glove" helps people who have problems moving the wrist and hand due to a stroke or spinal cord injury. As long as the person has some ability to move the wrist, the glove can help stimulate the muscles in the forearm that control grip. Sensors detect wrist angle and then trigger small stimulators to electrically activate the flexor muscles of the forearm. Imagine picking up a bottle of water. As you reach out and pick it up you first open your hand, make contact with the bottle, then close your hand, and lift it up. If you had partial paralysis of your arm and hand this would be difficult, if not impossible, to do. With the bionic glove, the user reaches out and gets the hand around the bottle. But because she cannot contract the flexor muscles, she extends the wrist a bit more. This signal triggers the glove to flex the fingers and the grip is made. Then the bottle can be picked up and used. In this simple device, three different muscle groups are activated with electrical stimulation.

So far our examples have had to do with neuroprosthetics that detect signals related to a residual movement that someone can make (like a bit of a walking movement, a bit of a wrist movement). Then the devices use that signal to trigger a stimulator to activate muscles that cannot get the normal activation from the nervous system. It is important to understand how these types of devices function to get closer to appreciating how the Iron Man suit could actually work. For starters, the suit would have sensors detecting nervous system commands from nerve or muscle as well as commands from residual movement. The suit then would amplify the normal movement. However, it wouldn't do so by stimulating the muscles like in FES. Instead the trigger signals would drive the motors controlling the joints in the Iron Man robotic suit. Above we were talking about restoring function in a damaged nervous system with FES and neuroprosthetics. In that way the neuroprosthetic helps "bridge" the problems in the nervous system to restore some movement ability. This is what Tony will need to operate the NTU-150 and already exists in the form of the Cyberdyne Hybrid Assistive Limb (HAL) wearable robot suit.

HAL is a kind of robot suit that is worn in order to improve physical capability. As we already learned, when someone tries to make a movement, weak electrical signals travel in the nerves and occur in the muscle during a contraction. These weak signals can be detected, measured, and amplified with electrode sensors placed on the skin over the muscles being used. The HAL suit uses this control signal to trigger the control of motors acting at joints on the suit. As a result, the suit is controlled directly based on the commands coming from the person wearing it. So, controllers for the elbow joint motors are triggered from nervous system commands going to the muscles that normally flex and extend the elbow. Cyberdyne Inc. likes to call this a "voluntary control system." This type of system relies on the users' intended movements to then amplify those movements by making the robot suit do the appropriate action. An additional layer of control is added using a "robotic autonomous control system," which is a kind of predictive system that works along with the voluntary triggering. All together, HAL applies a hybrid of the two control modes that provide an almost human-like movement. We will pay more visits to HAL later on in the book.

This basic concept of hybrid control has also been used by a company called Touch Bionics in their development of a fantastic neuroprosthetic hand. Think back to our medieval "Iron Hand" prosthetic shown in figure 2.4. Touch Bionics has created a sophisticated robot hand prosthetic that is driven by the normal muscle activation signals for the fingers. It can also be controlled by touch signals taken from pressure sensors. The Touch Bionics 5 finger i-LIMB hand uses inputs that come from the normal muscle signals to

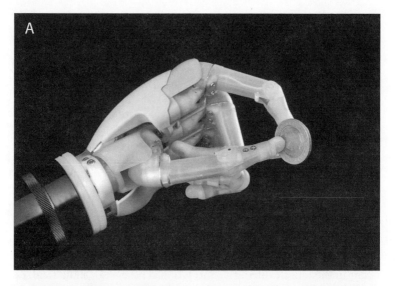

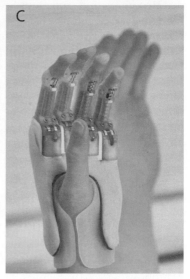

Figure 3.2. Touch Bionics 5 finger i-LIMB hand, which uses inputs that come from muscle signals to open and close the lifelike plastic fingers in the prosthetic. The i-LIMB makes a pinch grip (*A*) and individual "Pro-Digits" can be used for people with partial amputations (*B* and *C*). Courtesy Touch EMAS Ltd.

open and close the lifelike plastic fingers in the prosthetic. So, it uses the signals that come from muscles in the stump or remaining part of the person's arm. The i-LIMB hand then can open and close to grasp objects in a way similar to a biological hand (panel A of figure 3.2).

ProDigits is an application of this device for people who are missing one or more fingers due to accident or from birth. This device has individually powered and controlled motors for each finger and can be set up to take over for just the fingers needed by the user. This means that a lot more than just an open and closed grip can occur and more dexterous activities can be done, such as pointing with the index finger and typing on a keyboard. These seem pretty simple tasks—and they are if you have an intact hand. But they are not if you don't. An example of replacing one finger is shown in panel B of figure 3.2 and replacing function for four fingers is shown in panel C. The i-LIMB hand and the ProDigits can be covered in a flexible skin product making it look just like a real biological hand. Or they can be left uncovered. Tony Stark would go for the covered option if he needed one, I think.

Using Nervous System Commands to Control Iron Man

Now let's return to Tony Stark—someone with a fully intact body and nervous system—wearing a robotic suit to improve and amplify his normal abilities. If you think this through, you will realize that using the Iron Man suit could occur by tracking the nervous system commands and using them to control the suit. Doing this effectively takes the user's muscles out of the equation. That is, it creates the same disconnect between nervous system and movement that exists after a spinal cord injury or stroke. We just finished talking about using signals in the nervous system to trigger muscle activity and devices like robot suits or artificial limbs. The next step is determining the feasibility of using that initial command signal—the one from the brain or spinal cord—to power motors and computers directly. This means thinking about what Doc Ock from Spiderman or Professor X / Charles Xavier from X-Men can teach Iron Man about connecting machinery to his nervous system. What would it mean for Tony Stark to engineer the Iron Man armor to be able to use this kind of control? Is it even possible, and, if it is, is it dangerous? To answer this we are going to do a little fast forward and then a rewind!

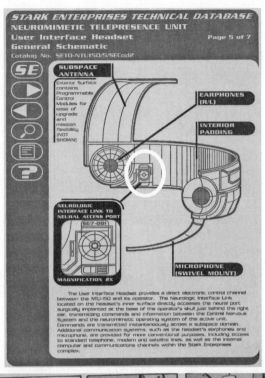

Figure 3.3. The "neuromimetic telepresence unit" that Tony uses to interface with his brain and to remotely control the Iron Man suit of armor (A), from the graphic novel *War Machine* (2008). Note the circled "neural access port" that is meant to penetrate Tony's skull. Tony connects to the telepresence unit (and therefore controls the Iron Man suit) from his hospital bed (B) from "This Year's Model" (Invincible Iron Man #290, 1993). Copyright Marvel Comics.

First, the fast-forward part. What we are focusing on here is the issue of somehow using a direct connection between the nervous system and a robotic device. This kind of connection was shown in Iron Man in its most extreme form back in March and April 1993 in "This Year's Model" (Iron Man #290) and "Judgement Day" (Iron Man #291). These stories contain elements of the extended story arc captured in the 2008 Iron Man graphic novel *War Machine* in which Tony Stark had to fake his death. Jim Rhodes has stepped in to become a fill-in "silver" Iron Man (and later became War Machine). Tony then has to use a remote control Iron Man (the NTU-150, but I will call it "robot Iron Man"), which is controlled by a direct connection to his nervous system called a "neuromimetic telepresence unit" (hence the name NTU). This unit basically involves a direct link between activity in Tony Stark's brain and activity in robot Iron Man. Included in the graphic novel is a detailed description of this telepresence unit.

The image shown in panel A of figure 3.3 comes from that manual. There is a lot of description in the seven-page pseudomanual printed in the novel! However, for our purposes, the piece I want to key on is the description of the actual headset the user must wear. It is of course called a "user interface headset" so the writers are taking a very literal view of how real scientists actually describe things! Anyway, as written in the manual, the headset "provides a direct electronic control channel" for the operator to use to control the robot Iron Man. This headset interfaces with the operator by "the neural port surgically implanted at the base of the operator's skull just behind the right ear, transmitting commands and information between the Central Nervous System and the neuromimetic operating system." The image in panel B shows the headset being interfaced ("jacked in") to Tony's brain and comes from "This Year's Model" (Iron Man #290). In both images, I have circled the key neural link panel. Sounds absolutely like comic book fiction, right? Well, partly it is but it also is very much like an emerging phenomenon generally known as a "brain computer interface." To explore this for Iron Man, let's look at the real science behind this concept.

Signals from Ol' Shellhead's Head

Since this section of the chapter is about detecting some information from the brain that is then relayed to robot Iron Man, we need to also understand how your own nervous system works to produce and regulate movement. That is, where does the signal for movement come from and what does it look like? I am pretty sure that you would agree with me that there are lots of things going on in your brain at any given moment. You probably also recognize that your brain doesn't exactly work like your computer or even your car engine. So, there isn't a little port just sitting there ready to have something plugged into it so it can directly relay commands to a computer. Is it even possible to get specific and useful information from all the activity in the brain? Let's investigate that "brain-computer interface" concept I mentioned a bit earlier. Get ready and hold on because we are about to dig deep into your gray matter.

Your brain contains about 100 billion neurons, and there are about 1 billion more living in your spinal cord. As I write this sentence there are about 7 billion people on earth. So, the number of neurons in your nervous system is about 15 times more than all the people on earth right now. If we think of activity of the neurons in the brain like individual people trying to talk to each other, we can ask ourselves this question: what—if anything—can we extract from a conversation among 101 billion people? Luckily all our neurons speak basically the same "language" and communicate in the realm of electrical signals. And, they don't all talk at once and aren't literally all connected to each other. Despite the fact that there are so many neurons with different levels of activity, amazingly we can get something consistent and resembling certain patterns.

Why is it that we can get anything to use as a signal to control things? When we make a purposeful movement, the commands start way up in our brains. Literally at the top, because the part of your brain that helps initiate movement really is at the physical top of your brain. (We will come back to this in more detail later in the chapter.) It is a bit oversimplified to say that areas of the brain are set up completely separate from other areas like isolated little kingdoms. However, different areas of the brain have very specialized functions, and it is usually shown as divided into frontal, parietal, occipital, and temporal lobes (also called "cortices"; figure 3.4). The labels in the figure within each lobe are meant to generally indicate the functions for those brain areas. When speaking about commands for movement, we are in the "motor system."

A common story in physiology and neuroscience is that many of the discoveries about function of parts of the brain and nervous sys-

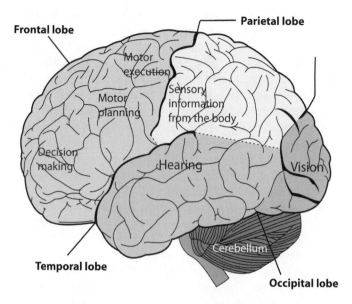

Figure 3.4. The human brain showing different areas of specialization in the cerebral cortex and the cerebellum. Modified from Mysid's adaptation of the 1918 edition of *Gray's Anatomy*.

tem have come from observing what happens when things don't work well or when there are injuries. In other words, much of what we knew before imaging technology came from descriptions of how movement control was disordered after brain or spinal cord injury. The "Edwin Smith Surgical Papyrus" described motor control problems after head injuries in ancient Egypt—over 5,000 years ago. Even though people have known about the connection between brain injuries and motor control for millennia, for quite some time there were many controversies about how the nervous system itself worked. For example, it took a long time to establish that the cells in the nervous system were "excitable tissue." That is, they convey information using electrical signaling (see chapter 2). This is very important for the issues involved with Inventing Iron Man, since many of the things we are discussing in this book have to do with interfacing electrical devices (like computers) with the basic signaling within the nervous system (which is electrical). However, in classical medicine, Galen

(AD 129–199) suggested that nerves were hollow and worked in a kind of pump or pipelike system to convey commands in the body. The substance relaying commands to activate muscle would then flow into the muscles and make them go. This idea was also favored by famous French philosopher René Descartes (1596–1650)—he of "I think therefore I am" fame. However, cutting to the heart of the matter (there is a pun intended as you will read), Alexander Monroe (1697–1762) showed that cutting a nerve did not reveal a gushing or outflowing from the nerve. This would have to have occurred if the older ideas of Galen were correct, so Monroe's experiment proved this wrong. Monroe thought maybe electricity might be involved instead.

This idea was met head on—with lots of controversy about "animal electricity"—by two very important Italian physiologists, Luigi Galvani (1737–98, from whose name we get galvanic current and the word "galvanize") and Alessandro Volta (1745–1825, from whose name we get volts as a measure of electrical amplitude). Galvani showed that a frog leg could twitch even (shortly) after death if the nerves going to the leg muscles were electrically stimulated. The controversy arose because Galvani thought this electrical stimulation used electricity within the frog's leg (e.g., animal electricity), whereas Volta thought that the frog's leg was merely a conductor of electricity. So, the combined research of the two men was the first real description of the electrical nature of the nervous system. However (this bit is really important, so pay attention please), when the brains of different animals were stimulated with electricity, not much actually happened. This suggested that maybe the brain didn't do anything specific and related to the control of movement.

In fact, Charlotte Taylor and Charles Gross have described how, up until the eighteenth century, the outer surface of the brain (known as the cortex) was actually considered to be a useless "rind." By the way, this is actually what the root word "cortex" means in Latin. Some scientists correctly disagreed. Thomas Willis (1621–75), a professor at Oxford, and Francois Pourfour du Petit (1664–1741), a surgeon in the French army, both thought the cortex had an important role in movement control. In particular, from observing lesions in injured soldiers and from parallel experiments in dogs, du Petit noted that the outer surface of the brain was indeed very important for movement. These observations from hundreds of years ago helped show that the brain and nervous system were electrical in nature and that there were specialized parts of the brain, including those related to movement.

Clear evidence of specific functions in different parts of the brain had to wait until the excellent work of Paul Broca (1824–80). In 1861 he wrote about several patients who had difficulties in speaking. They all had damage to the left frontal lobes. This showed clearly that certain functions (in this case, speech) could be largely controlled and affected by very specific parts of the brain. You can roughly locate this part of your own brain by running your hand over the corners of your forehead as your skull moves back toward your ear. Anyway, it would take a bit more creative work after Broca's research to convince people that parts of the brain participated in movement control.

It is often said that the human brain is the most complex organ. Measuring activity in such a complex organ is not as simple as you might imagine. Remember, there are 101 billion neurons to listen in on. And they have to communicate together in useful patterns in order to produce all the behaviors we are capable of. Technology has often been a limitation for this kind of measurement and only small numbers of neurons have been recorded. In 2007, MIT neuroscientists Timothy Buschman and Earl Miller conducted a study aimed at looking at attentional focus in monkeys. They recorded from up to five hundred neurons simultaneously in three different brain regions during different tasks of focusing on targets. This represented a huge advance in the ability to record a large numbers of neurons simultaneously.

Creating Commands from the Cortex

An important insight into the role of the cortex in movement control came from the work of John Hughlings Jackson (1835–1911). He was a British neurologist who studied patients with epilepsy. His clinical observations suggested to him that certain parts of the brain must be closely related to specific motor commands. He saw that during a seizure there was a consistent and organized spread of muscle contraction across the body. This made him think that certain parts of the brain should have specific actions in causing movement and that the whole system must be organized in a way reminiscent of the layout of the body. However, he had to wait until the work of Gustav Fritsch (1838–1927) and Eduard Hitzig (1838–1907) for confirmation. Fritsch, while working as a military surgeon had noticed that his efforts to treat a head wound would sometimes (accidentally) cause small contractions on the side of the body opposite to the injury. In 1870 Fritsch and Hitzig used electrical stimulation of the brain to generate detailed maps of the brain of the dog and showed clearly that movements could be created by stimulating certain brain areas. So, at this point it was known that electrical stimulation of certain parts of the brain (but not others) could evoke twitches in muscles of the body and that there was a kind of map of the body muscles represented somehow by the neurons in the brain. These studies also revealed that the control of activity in muscles is generally found on the opposite side of the brain. If you are using your right hand to turn the pages of this book, it is the cells in the motor cortex of the left side of your brain that are sending the commands. Also, if you choose to turn the page with your left hand, it's the command cells in the motor cortex on the right side.

Canadian neurologist Wilder Penfield and his friend Edwin Boldrey followed up on this work of locating the centers for different functions in the human brain. They did a detailed stimulation exercise and found that they could generate a kind of "map" of the muscles of the body from stimulation of the brain. The basic concept is this: if you give electrical pulses of stimulation to the motor areas of the brain, you can trigger the output cells of the brain to relay commands to the cells in the spinal cord that activate muscle. By moving electrodes over the surface of the brain, movements in different muscles can be observed. Through painstaking effort, it is possible to create a kind of input-output map of the surface of the brain, which is weighted differently depending upon how much area (and therefore numbers of cells) on the brain are devoted to a particular part of the body. Think of how a huge city with 15 highway interchanges compares with a small village with no exits off a highway are represented on a road map.

Penfield and Boldrey's work was the basis for the "homunculus" (little man) concept that describes the map of the muscles of the body on the surface of the brain (figure 3.5). The surface area of the body on the map is an indication of the number of brain cells controlling those muscles. These cells are found in the "motor execution" part of the brain shown in figure 3.4. We also have similar maps related to the sensory areas of the brain. In that case, the maps are created by recording activity of brain cells when different skin areas are activated. Understanding this is important in grasping whether it might

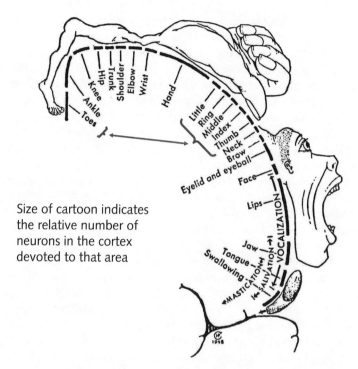

Size of cartoon indicates the relative number of neurons in the cortex devoted to that area

Figure 3.5. "Map" of the neurons (upper motor neurons) in the brain used for activating muscle. The distorted shapes of the body part represent the relative number of neurons that control muscles in that part of the body. Modified from Penfield and Rasmussen (1950).

be possible to tap into this system to control computers and robotic devices. To set the stage for that, I think it is probably useful to ensure that we understand how movement commands arise and are relayed.

Peeling off the Shell

Now let's press ahead and look at how the signal for the activation of Tony Stark's muscles actually occurs. To begin, you have activity in the brain (there are a number of places where this occurs, actually, and we will come back to this shortly). We will focus right now on that part of the brain where we find the motor map we were just discussing. Activity from these cells is sent down to the spinal cord in the form of what are known as "action potentials" and then out to the muscles. Recall in chapter 2 we learned about the movement of sodium, calcium, and potassium ions in and out of cells and how this was linked to the electrical energy needed to move muscles. An action potential results when this energy rapidly rises and falls.

What if you decide you want to make a movement? It could be a motion as simple as picking this book up from the bookstore shelf. Or it could be as complex as trying to jump into the air and fly, in true Iron Man fashion. Let's say it was the former, because we can deal with that directly. Also, I don't support you actually jumping into the air and attempting to fly around. As soon as you decided to pick up this book, a command moved through several relays and then finally arrived at the main movement output center of your brain—your motor cortex. This is the home of the neurons that send along the command to activate muscle down into the spinal cord. This part of the motor cortex is right beneath about the topmost part of your skull. You can get a rough idea of where it is by finding the top of your ear and running your fingers up over the topmost middle of the skull and then down the other side. The cells in the motor cortex found under your fingers relay the motor command and are known as the upper motor neurons. They are called this because these motor output neurons are at the physically highest ("upper") part of the nervous system. After that, the relayed command to move arrives at the spinal cord level (these are the lower motor neurons) for the appropriate muscles.

So that is how you and I and Tony Stark deliberately contract our muscles. But that is just the direct output part. What we also want to understand is how we can detect that activity in the brain related to activating the muscles, but then kind of "short circuit" it so that we can use that brain command to activate a computer or a robot or a motor. Or maybe a computer-controlled, motorized robot (can you say Iron Man?). In so doing, we will answer the question of where the commands to start movement actually come from.

Who Makes the Plan?

Two other parts of the brain play important roles in the control of movement. These areas are specifically involved in the planning and coordination of movement and go by the clever names of premotor and supplementary motor areas. Inside your skull, these two areas would be found by locating the middle of your skull as for the motor cortex. Then move your fingers forward about the thickness of two or three fingers and you are into regions right beside the primary motor cortex.

To tackle the question of how we could interface with the brain in order to control machines, we turn to what and how can we get information from brain activity. This brings us to the concept of recording activity from the brain, so our next stop is to understand a little bit about electroencephalography, also known by its initials of EEG. We spoke earlier about Galvani, Volta, and electricity in the nervous system. Here we are talking now about electrical activity in the brain. The activity of all those neurons generates electrical field potentials that can be measured by putting electrodes over the scalp. These noninvasive measures were first discovered by German scientist Hans Berger in 1929, but the concept of electrical activity in the brain was originally described by Englishman Richard Caton in 1875. The thing is that the brain activity, as taken from the EEG signal, changes depending on what you are doing. The size of the activity changes as well as the number of "spikes" that you can see. All of these represent changes in the overall activity of neurons in different parts of the brain. Although it gets a bit complicated, this EEG activity can be filtered and analyzed and then used as a control signal to affect computers and robotic devices. This is called a brain-computer interface and brings us back to the telepresence unit that Tony Stark created for the Iron Man armor. So this part of the Iron Man mythology is already a reality.

You can appreciate the input and output of the brain by actually stimulating the neurons to make them become active. Electrical stimulation over the scalp or on the brain surface can be used. Or, a common research technique (and one that is now used clinically too) is to apply transcranial magnetic stimulation. Conveniently, electrical and magnetic fields are interchangeable, and we can use a magnetic field to activate electrical neurons. This involves using a powerful magnetic coil placed over the part of the brain containing the neurons controlling the muscles you're interested in.

Figure 3.6 shows me sitting in a chair with a magnetic coil placed over my scalp on the left side of my head. Since the pathways for the motor output cross over to the other side of the brain stem and spinal cord, the cells in the left cortex control the muscles on the right side and vice versa. If I were to make a slowly increasing contraction with my forearm flexor muscles (the ones that pull my wrist in), I would slowly increase muscle activation and force production at the wrist. We can mimic this by steadily increasing the stimulation intensity. Three examples of different intensities of stimulation are shown at the bottom of the figure. You can see how the response of the muscles (called "motor evoked potentials," or MEPs, and measured with electrodes over those muscles) increases as stimulator output goes up. This shows the clear relation of input and output. We could also basically do the opposite. Instead of stimulating the motor cortex and recording EMG in the muscles, we could record EEG activity from the somatosensory cortex while we stimulated the skin on a body part, resulting in a "somatosensory evoked potential," or SEP.

In clinical neuroscience, tapping into brain commands for movement has been used to try to help people with certain neurological diseases that affect movement. The terrible disease amyotrophic lateral sclerosis (ALS, also known as Lou Gehrig's disease) is one example. In ALS the lower motoneurons in the spinal cord all progressively die. During this process, the person becomes steadily weaker, is eventually paralyzed, and survives only until the motoneurons controlling the muscles for breathing die. It is one of the most terrifying neurological diseases in my view. The remarkable thing about ALS, though, is that the upper motoneurons, the ones up in the brain that we were discussing earlier, do not die. In fact, ALS doesn't affect any of the cells of the brain. In an attempt to help moderate the effects of the disease, scientists and clinicians can train people who are early into their ALS to use the EEG signal to control a computer cursor.

Making a simple device such as this available to many has been the passion of physician and neuroscientist Jon Wolpaw. Since the early 1990s, Jon and his team at the Wadsworth Center of the New York State Department of Health in Albany have been working on developing a brain-machine interface system based on EEG brain wave activity recorded from the scalp. Recently, this group has developed a brain-machine interface that can be taken into the homes of users. The Wadsworth device is used by people paralyzed in end-stage ALS, and other neurological disorders in which motor control is lost, to communicate by measuring brain activity with a simple stretched cap containing electrodes (embedded into the cap used in clinical EEGs).

Transcranial magnetic stimulation of my left
motor cortex controlling my right forearm muscles

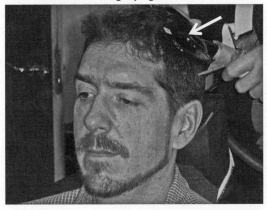

Activity in my right forearm muscles

Increasing output
from my brain
with magnetic
stimulation

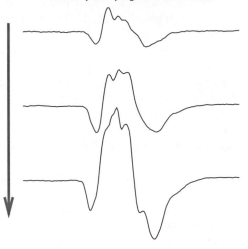

Figure 3.6. A transcranial magnetic brain stimulator activating the
neurons in my brain that control my right wrist. Each stimulation
caused an involuntary twitch of my wrist muscles. This twitch got
larger (like trying to contract more forcefully) when the stimulator
was turned up higher. Courtesy Richard Carson.

The interface system is used to measure brain activity while the person observes a computer display with items that relate to a standard PC keyboard. The interface then determines which keyboard item the person wants to use. This system can be used to write e-mails and operate any PC Windows-based software that can be controlled by keyboard interface. Currently, this system still requires ongoing intervention and monitoring by experienced support staff, who must come to the user's home and also remotely monitor activity. The Wadsworth group currently focuses much of their efforts on trying to minimize this need for costly technical support.

Currently, some people with ALS who are approaching complete paralysis are using the Wadsworth brain-machine interface. They were able to control a computer interface that allowed them to move a cursor on a screen to select letters to spell words. The idea was that it would be helpful when the disease progressed to the point that they couldn't speak. So they trained the participants how to use the devices when they still had some use of their limbs and then they were well placed to be able to use them in the late stages of the disease.

At the other end of the spectrum, several toy companies are coming up with similar devices with video game controllers. Mattel Mind Flex, NeuroSky, and Emotiv all use scalp electrodes to detect EEG activity to move cursors in games.

Brain-Machine Interfaces Put Thoughts into Action

The basic concept of a brain-machine interface is essentially replacing the biological signaling connections with technological ones. When damage occurs in the nervous system, such as after a stroke or spinal cord injury, there is interruption in the normal signaling connections from brain to spinal cord that leads to difficulty in muscle activation. As an example of the effect of trauma, let's consider someone who experienced a spinal cord injury in the neck.

The late Christopher Reeve (1952–2004) experienced a horrific spinal cord injury when he was thrown off a horse he was riding. He shattered two vertebrae (the bones of your spinal column) just below his head at the top of the neck. These were cervical vertebrae 1 and 2 (going from top to bottom you have seven cervical, 12 thoracic, and five lumbar vertebrae). Spinal cord injuries are graded in severity based on the level of the injury (higher is worse because more "downstream" parts of the spinal cord are affected) and how "complete" it is, basically on how damaged is the spinal cord. An injury at the C1-C2 level is often fatal, because it affects the parts of the brainstem that control breathing and cardiovascular function. Christopher Reeve is a tragic example, but he is a good one to think about in a book about a superhero since he played Superman in four major motion pictures from 1978–1987. After his accident, he was required to use a ventilator to breathe and had no functional ability to activate any arm or leg muscles.

If a fully developed brain-machine interface had existed, it could have been used to detect motor signals in Christopher's brain that signaled his intention to pick up an object. Perhaps a glass of something. Or a coffee mug. That would be a typical textbook example. However, I want to use the example of a New York Rangers hockey jersey—I will explain why in a minute. Using a brain-machine interface and a robotic arm, the command to pick up the jersey could be relayed to a computer controlling the robotic arm, and the controller would bring the jersey close to Christopher. As a point of reference, we are nowhere near having anything this complex at present—although researchers have been able to get a monkey to feed itself an orange using this kind of system. An ideal interface would—with no obvious delay or difficulty—take the thought about the action and transform it into actual action.

Now, let me briefly come back to the reason for the New York Rangers jersey. I had the good fortune to meet Christopher Reeve in 2001 at an international spinal cord injury research conference held in Montreal, Quebec. He told us how much he liked the city of Montreal and how, in 1986, during the filming of *Switching Channels*, he wore a New York Rangers jersey to an NHL playoff hockey game between his Rangers and the Montreal Canadiens. Unfortunately for him, but fortunately for legions of Canadiens fans (such as me), the Rangers ran into a very hot, future hall of fame goalie named Patrick Roy and lost. Christopher explained his passion for hockey and his enjoyment of watching playoff hockey between New York and Montreal (two of the "original six" founding members of the NHL). He was presented with a Montreal Canadiens jersey by the conference organizers, which was what spurred the story I just related. So there.

An interesting example of brain-machine interface using implantable electrodes is the CyberKinetics BrainGate. BrainGate's mission is stated as "advancing technological interfaces in order to help neurologically impaired people continue to communicate with others." The objective appears to extend to activities including the control of objects in the environment such using a telephone, television, or room lights. The basic BrainGate system includes an electrode sensor that is implanted into the motor cortex and connected

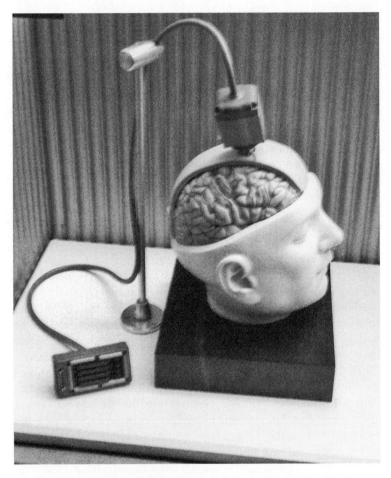

Figure 3.7. Electrode arrays implanted in the brain using the "Brain-Gate" system. Courtesy Paul Wicks.

to a computer interface that analyzes the recorded neuronal activity. The system simultaneously records electrical activity of many individual neurons using a silicon array about the size of a baby aspirin. That array contains one hundred electrode contacts that are each a bit thinner than a strand of your hair. Figure 3.7 shows an anatomical model of the head with the electrode array inserted into the brain through a port in the skull. The other end of the cable goes to an interface cable that can go to a computer. The principle of operation is that the neuronal signals from the brain are interpreted and translated into cursor movements. This means the person can control a computer with thought, in a way similar to using wrist movement to shift a mouse to move a cursor. This is close to the concept of the NTU-150 telepresence armor that Tony created way back in 1993. That's where (when?) we go next.

Brain-Machine Interface and the Iron Man Neuromimetic Telepresence Unit

Let's return for a minute to Iron Man's telepresence armor. Recall that the telepresence unit responds to control from the user. The main outline of how this is supposed to work is shown in the "Stark Enterprises Technical Database" and the "communication network schematic" in the 2008 *War Machine* graphic novel series. The graphic novel depicts a flow chart linking the robotic remote controlled armor and the headset that Tony wore in "This Year's Model" in Iron Man #190. The caption for it reads: "The diagram below represents the basic operation of information transfer between the User Interface Headset and the NTU-150. Because the actual process incorporates many hundreds of individual system checks, security interlock codes and neurological failsafe routines, the chart has been simplified to display only the primary system events." So, now you know why it is such a streamlined schematic! There are indeed some great similarities between how the Iron Man NTU-150 system and a basic brain-machine interface works. Essentially both involve extracting information about movement from brain activity, which is then processed into a command to control a device. By far the most complex behavior demonstrated to date has been monkeys learning to feed themselves oranges using a robotic arm controlled by brain activity. Human studies have not reached this level of sophistication, even with brain electrode systems. So, we are a good way off from being able to remotely control a robotic suit of armor with brain-derived signals!

Another interesting tweak for the Iron Man system that is not yet practicable in real life is the information flow shown in the communication schematic, which refers to information coming back from the device ("from NTU-150"). This kind of "closed loop" (where sensation feeds back into the system) would be an absolute requirement for telepresence armor and similar devices to work in practice but is considerably far away from being implemented.

However, some recent work in a related area may one day pave the way for this kind of system. Deep brain stimulation is a technique that involves implanting electrodes into the base of the brain, usually into parts of the basal ganglia and into parts that are important for controlling movement. These areas of the brain are the main ones affected by the progressive neurological disorder of Parkinson's disease. To help with the difficulties in producing movement and the tremors that are very common in this condition, many treatments are used, including taking certain drugs that affect dopamine systems. Current research into deep brain stimulation changes the activity in this part of the brain, and it can be really helpful in improving movement. The procedure involves setting the stimulator externally and then observing the effect on the user. Any changes in the stimulation have to be set externally in what is known as "open loop" control. A better and more adaptable system would be to have closed loop control, which is essentially the way the Iron Man NTU-150 is likely meant to work. And up until very recently no deep brain stimulator included this concept. The Medtronic Neuromodulation Technology Research division has developed a preliminary system that can extract information on brain activity that can then be used to change the settings of the stimulator. Rather coolly, Stanslaski and colleagues reported that this sensing system is rather BASIC. As in Brain Activity Sensing Interfacing Computer. While this is still a long way from the NTU-150 telepresence armor, it is actually a direct step along a path that is heading in that direction.

Related to these steps is the concept of an optical neuroprosthesis. Illustrated in figure 3.8 is a summary of different approaches to supplementing vision in visually impaired people. These images come from the work of Eduardo Fernandez and colleagues and represent a fascinating parallel of work others have done with cochlear implants to restore hearing. At the top of the figure, three different

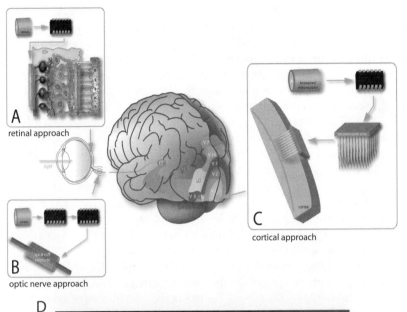

A
retinal approach

B
optic nerve approach

C
cortical approach

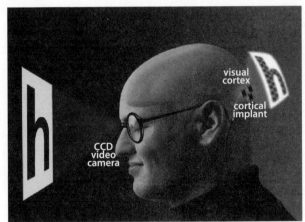

D

Figure 3.8. Visual neuroprosthetic interfaces. Panels A–C show different approaches and locations for "tapping" in to the flow of information in the visual system. The bottom image shows an approach that uses video camera inputs from glasses, which then activate the visual cortex of the brain. Courtesy Fernandez et al. (2005).

"approaches" are outlined. Normally, visual information flows from the retina via ganglion cells into the optic nerve and eventually to visual cortex in the occipital lobe. Panel A shows the idea of using a neuroprosthetic eye to connect with neurons (ganglion cells) in the retina. Panel B uses the approach of directly activating the optic nerve (which contains the output from ganglion cells), and panel C shows the concept of connecting directly to visual cortex. This idea is highlighted in the bottom panel where the most "high-tech" approach (at least in appearance) is shown. In this case a camera in the lens of the glasses takes visual information and, after processing, feeds directly into the visual cortex by way of the cortical implant shown at the back of the head. This would be like taking the visor information from Iron Man and feeding it directly into Tony Stark's brain. Or, as an example from *Star Trek: The Next Generation*, of using Geordi's visor to send visual information to his brain. This is staggering stuff and exploration in this field continues at a rapid pace.

I have been dwelling on this system and example in Iron Man for so long because it gets to the heart of whether Tony could really become Iron Man: the robotic control of the suit. Instead of a remote-controlled suit of armor, Tony would have to use a brain-machine interface to control the suit. But how difficult would that suit be to control? And would your body like it? We are going to answer those questions in the next chapter.

The First Decades of Iron

"He Lives! He Walks! He Conquers!"

Just developing and learning how to use the Iron Man suit would take up the first five years of a journey to invent an Iron Man. The technology to develop an armor system with articulated armor currently exists. Tony Stark—or anyone else who wants to follow in his footsteps—would need about two years to adapt such technology to create the full body armor that we see with Iron Man. An additional four years would likely be required to strengthen and lighten the suit and then incorporate it all into a fully mobile passive system. Such a system would make it move like a high-tech suit of armor reminiscent of medieval knights but with much more freedom of movement.

To provide the "extras" that Iron Man needs for his life of crime fighting would require motorizing and energizing the armor. A prototype with these features would take another four years. Even when this improved armor was complete, its user would encounter a major problem: the current standard for this kind of approach in industry is to use hydraulic actuators (think forklift), which are extremely heavy. So Tony would still have to focus on making the armor even lighter and the motors much smaller and efficient for the future.

A key focus for Tony in the next ten years of developing his suit would be to miniaturize the motors that control the movements of the fully powered armor. This would include getting away from hydraulics. Although research ideas are currently under way for this type of refinement in real life, progress is slow and the technology does not yet exist. Tony would have to invent the new type of motor himself, which would likely take five to seven years of work. A stumbling block for full implementation of this improved suit here is the need to power the armor for independent movement (that is, while away from a fixed power source), which would require development of new power cell technology. It would also include the harvesting of energy from the movements themselves, and it's uncertain how long that would take.

Tony would have to incorporate into the armor a movement-triggered control of the motors at this stage. That is, triggering motor control in the exoskeleton by small movements made by the user. This currently exists for the extremities—hands, feet—and simple movements—grasping, walking—but would need to be fully integrated into a whole body armor system. This would take an additional two to three years. After completing all of this, Tony's biggest challenges still await him.

PART II USE IT AND LOSE IT

Will time tarnish the Golden Avenger?

Multitasking and the Metal Man

HOW MUCH CAN IRON MAN'S MIND MANAGE?

My armor has seven advanced genocide mechanics troops tracked and targeted. It's relaying suit performance data back to Pepper on the helicarrier. It's keeping an eye on a communications satellite over Madrid that's either being hacked or starting to fail. It's relaying a PowerPoint presentation from a Stark U.K. R&D presentation. And apparently Josh Beckett is eight innings into a no-hitter . . . Not to jinx anything.
—Tony reflecting on all the info the suit provides, from "The Five Nightmares, Part 2: Murder Inc." (Invincible Iron Man #2, 2008)

Outside, a 400 mph slipstream of freezing air is roaring past me at a sound level of 104 decibels. Inside, a 9,000 song playlist that's heavily skewed to '80s metal is roaring at only a few decibels less. On my back, a superconducting capacitor ring is spinning, charged with enough electricity to power a decade-long concert by every band on that playlist at once.
—Tony Stark, from "Hypervelocity: Part 5" (Iron Man, 2007)

Y ou are making dinner and just added pasta to boiling water. You have a cup of hot tea in one hand. Then the phone rings so you run over to grab it. At the same time your dog starts barking to be let in. Or out. Or maybe she isn't sure. But you let her out while answering the phone. Just in time you glance over at the stove to see the water foaming up and about to boil over. You wedge the phone under your chin so you can fling the door open (or closed) with the hand not holding the mug and run over to the stove. On the way you don't notice the dog's squeaky toy, trip over it, stumble, spill your tea all over the work you left on the table and arrive at the stove just after the water has boiled over leaving a nice white scum that you will have to clean up later. You just did a lot of multitasking. And it didn't all work out.

Most of the time, it is fairly simple to perform a wide range of movements or tasks. We seem to sometimes perform even more specialized skills like driving with little obvious attention. In our society, we now do a great deal of multitasking, and juggling many tasks all at once is commonplace. However, when we have to do different things simultaneously and as the need for skill and complexity increases, tasks become more difficult. The scenario we just opened with is a good example. You can probably call this the "walking and chewing gum at the same time" problem. Imagine walking across a room (or, if you are able, you can actually do it). Now get a glass and fill it right to the brim with water. Hold that glass in your hand and then try to walk across the room again, all the while focusing on not spilling a drop of water. Probably when you did it that way, you either walked slower or walked at the same pace but spilled a fair bit of water. This outcome represents the effect of "cognitive load," which means that we can only put attention on so many things at once. The more we add to what we are doing, the greater is the degradation of performance of each thing that we attempt to do.

In this chapter, we will explore this specific problem. We will also look into what has been done to minimize the effects of cognitive load in real-life jobs that share some of the same concerns as Iron Man—fighter jet pilots. This chapter is about the limits of human information processing, or what we would more commonly think of as attention. And this is something that Tony Stark recognizes as an issue. In the story "World's Most Wanted, Part 6: Some King of the World" (Iron Man #13, 2009), Tony says, "The Iron Man is getting more complicated than I can pilot. I need to downgrade it back into something more . . . consumer grade." Later in that same story, he echoes this sentiment by saying "I have to simplify the suit." Let's look at why he has to simplify that suit!

How Much Does It Cost to Pay Attention?

Linking up a human body with technology has its limitations. A reasonable place to start is with one of the most commonly seen pieces of technology—the cell phone. Love 'em or hate 'em, almost all of us have used a cell phone and we have certainly seen many, many cell phone users. Just based on cellular subscriptions, almost 90% of the U.S. population uses cell phones. What I want to focus on here isn't just to do with using a cell phone, but rather using a cell phone while doing something else. Particularly, how much attention does it take up to use a cell phone and should you use one while driving a car? Or, even better, a mechanical suit of high-tech armor? Just this morning while driving to work I was stuck behind an SUV at a light with an advanced green signal flashing. Where I live, drivers who are paying attention realize that the advanced green signal means we get to turn left. However, the driver ahead of me failed to notice the signal. When I looked closely, I realized that the driver was instead talking on a cell phone and not really paying attention to what was happening. Why does that happen? To answer that means asking how much attention do you have to play with and are all tasks created equal?

Is it really a problem to use a cell phone while driving, or was I just a little put off because the driver kept me from turning left? Only if the answer to the first question is yes is it relevant to figure out why. It has been estimated by Marla De Jong that 85% of cell phone users in the United States use their phones while driving. Other studies have shown that cell phone use can increase the risk of a crash during driving fourfold. So, clearly, it is more than just a guess that using a cell phone while driving impairs driving performance. Why? Part of this has to do with the attention demands of speech and the idea of "inattention blindness."

Speaking and talking are motor acts that involve listening and attention. It turns out that listening takes up less of the activity in our brains than does speaking and getting ready to speak. The interesting part of all of this is that when we are on a phone listening to someone, preparing to speak to them, and then actually speaking we are constantly trying to figure out where the person is. A kind of mental image of where the person is and who they are forms. Maintaining this takes up a lot of the processing power in our brains. It is almost like the ancient part of our brain is constantly searching for who we are talking to. But it cannot "see" them. "What is that voice in the air?" we might ask in "Cirroc" (aka Phil Hartman's) voice right out of *Saturday Night Live*'s "Unfrozen Caveman Lawyer" fame. This is likely why just listening to music or talking to someone in the same vehicle does not cause the same distraction as talking on a cell phone. By the way, this applies to hands-free phones as well as normal cell phones. The difference is that with the hands-free phone the motor act of holding the phone is gone. This makes it marginally better but still doesn't address the issue of attention.

It should go without saying (but I will say it anyway) that this problem is even more dramatic when the cognitive task is combined with a motor act, like texting and driving. This actually does happen and that can have horrific outcomes. Tragically, on July 24, 2009, a driver ran her car into the back of a public works truck in Edmonton, Alberta. A city worker who was collecting pylons from the road behind the truck was crushed to death. Witnesses reported that the driver of the car did not apply the brakes at any time and even exited her vehicle after the accident while continuing to text. The driver pled guilty to a charge of careless driving. Many jurisdictions have now moved to ban handheld cell phone use (and texting by default) while driving. Distraction during multitasking with technology even as "simple" as an automobile and a phone don't mix. Let alone considering the implications for piloting Iron Man.

The ability to multitask also changes with age. Most people realize that older adults have an increased chance of falling when standing or walking. Many changes in the body make this come about, but some of it has to do with a reduced ability to control the body moving in the environment. That is, to manage all the information available. So, if someone has to walk and pay attention to the terrain—like walking down a staircase—carefully and think of something else they are managing many different pieces of information at once. This is often explored in scientific research using what is called a "dual task" paradigm.

I mentioned above the expression of "walking and chewing gum at the same time." Maybe a better one is to try rubbing your stomach.

Then try tapping your head with the other hand. Now try doing them at the same time. You likely noticed that you actually didn't have to pay much attention to either rubbing or tapping but when you did them at the same time you really had to think about it. Especially at the start. If you really want to make it harder, try to tap your left or right foot as well. Your ability to manage many tasks is limited and goes down as you age. So, in the dual task paradigm, older adults who have to do more than one thing wind up doing both things poorly. How does that relate to Iron Man? Running around as Iron Man involves a lot more than a "dual task"! I suggest it would be more like "centi-tasking"—doing a hundred things at once.

But, as with so many other things, a lot of a person's ability to multitask has to do directly with his or her level of training. That is, you cannot just jump into the Iron Man suit of armor (or any technology) and use it without training. It turns out that even the ability to avoid or reduce attention conflict in using a cell phone can be trained. James Hunton from Bentley College and Jacob Rose from Lincoln University did a neat study to look at this. They used a "simulated" driving course and had people have no conversation, a real conversation with a passenger, and a real conversation over a hands-free cell phone. They were keen to go beyond the small attentional issue of actually holding a phone because the biggest problem for attention is the conversation with someone who isn't physically in the line of sight. As you might guess from the other things we talked about, they found that even the hands-free cellular phone call increased the number of driving incidents and the number of crashes drivers had. By the way, a conversation with a real passenger led to increased problems also, but nowhere near the cell phone level. The extra twist was that they compared people who had pilot training with those who didn't have training as pilots. The pilots did better, much better actually, than ordinary folks. It isn't clear yet if those pilots got better at multitasking by the practice that occurred in training to be a pilot, or if they were better multitaskers to start. Pilots also expressed less desire to "visualize" or "see" the people they were talking to than did the nonpilots.

So far this kind of research suggests a surprising result. A major issue is the need to use some of the brain's limited attention to try to imagine the person being spoken to. This, combined with the fact that, unlike a passenger in the vehicle who can see what is going on, people on cell phones cannot modify their patterns of speech and

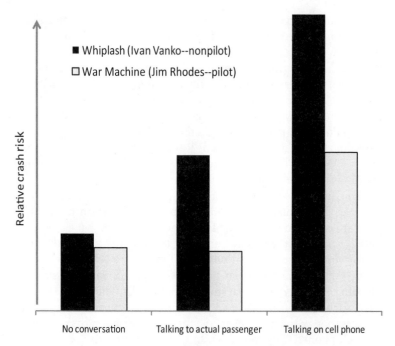

Figure 4.1. Comparison of the effects of multitasking and driving crashes when having no conversation, talking to a person in a vehicle, or talking on a cell phone for pilots (e.g., Jim Rhodes) and nonpilots (e.g., Ivan Vanko). Data redrawn from Hunton and Rose (2005).

conversation. Together, these impediments to attention are a dangerous mix and threaten necessary attention for driving safely. I have replotted some of Hunton and Rose's data and put an example together in figure 4.1. For contrast, I have used Jim Rhodes as an example of a pilot and Ivan Vanko (aka "Whiplash") in *Iron Man 2* as a nonpilot. Bottom line? Rhodey can talk to Jarvis or Pepper Potts while flying and won't have much risk of increased crash. Whiplash should try the strong silent routine at all times, even if he co-opts a suit to fly around in. Which, as we saw in Mickey Rourke's film portrayal of him, he mostly did.

Another study done at the University of Utah by James Watson and David Strayer also looked into "dual task" interference. They had people in a driving simulator (main task) and then use a cell phone (secondary task). To make sure anything they found was really about the interaction between tasks, they also had the participants use a hands-free cell phone. The main thing of interest was whether using the cell phone interfered with the reaction time to braking when a "driver" in the car ahead applied the brakes. As you might expect, a whopping 97% of the people studied showed a significant effect of using a cell phone and responded slower to applying the brakes when the person ahead of them did so abruptly. What was truly striking about this project, though, was that just under 3% of participants showed no performance interference at all. The authors called these people "supertaskers" and found that they represent a very small proportion of people who were able to excel at multitasking. (I find it very pleasing to use a word like "supertasker" when writing a book about a superhero.) It remains to be seen if this kind of "supertasking" ability can be gained or improved by experience or training. Clearly, to have any hope at all at using the Iron Man suit of armor, Tony Stark (and also Jim Rhodes as War Machine) would have to belong to this very small group of people.

The whole idea of attention is interesting all by itself. Lots of recent research has looked at brain activation during tasks that had a timing component as well as a location in space. Different brain areas get activated depending on what we are doing and what the expectation or timing is. Partly this relates to what would be called the perception of time. You might think of this as the "watched pot never boils" idea. How much attention we put on something can alter our perception of time. This, then, would be another wildcard for Tony Stark to deal with while he tries to maneuver the Iron Man armor.

Is Tony Stark the Pilot Made or Born?

Another question to think about is the extent to which multitasking in stressful environments can actually be learned or trained as opposed to being an inherent part of a person. Let's take the most complex "worst case" scenario for Iron Man. Despite the fact that we already recognize that flying around as Iron Man is not currently possible, let's still use that as an example of the control problem. Comic books and the movies from Marvel Studios often depict Iron Man as a kind of flying-suit-wearing jet-fighter. So, let's use the complexity, information management, and multitasking involved in military aviation as our real-life example of multitasking pilots and their ability to concentrate.

Recall the concept of embodiment we talked about earlier in the book on the topics of prosthetic limbs and of the Iron Man suit. Perhaps pilots learn to "embody" the aircraft to such an extent that they can dissociate the strangeness of multitasking and conversing remotely. The idea of pilots and jet fighters is a good reference point for Iron Man in action and his allocation of attention. The key here is imagining having to pilot the Iron Man suit of armor and having to deal not just with the tremendous attention even that would take up but also to be involved in what are essentially military operations at the same time. The kind of high-stress situation that we are talking about is when immediate and extreme negative outcomes can occur from the smallest error, a "kill or be killed" work environment for Iron Man. At the best of times, putting all of this together means that even ordinary and habitually practiced movements and skills can be difficult to perform properly and horrific mistakes can occur.

Related to the issue of cognitive load and distraction—and cell phone use—are some worrying statistics for flying. The U.S. Federal Aviation Authority reported in 2010 that of all the serious flying "events" (some of which involved crashes and fatalities) over half included violations of pilots and copilots having excessive conversations or using cell phones—being distracted. These are examples from the relatively peaceful concept of flying a civil aircraft. When we come back to Iron Man, it should be pretty obvious that the "task" of piloting a robotic suit combined with performing military operations would be huge. And largely impossible, if the suit of armor were something to be used like a tool and separate from the body of the user.

The main point of discussing these examples is to make clear just how much human performance and judgment degrade when in an extremely stressful environment or when distracted. We need also to consider the arousal level related to the level of stress. In psychology, this has been described as the "inverted U" (figure 4.2). There is an optimal level of arousal due to stress that allows for the best performance. In this context, "arousal" refers to heightened sense perceptions and mental alertness, both essential attributes for high-level functioning. Tony Stark as Iron Man needs some stress to have enough arousal to perform. And, since he has been successfully trained as Iron Man, he can perform at a very high level under stress (high

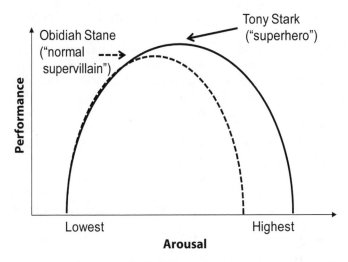

Figure 4.2. The "inverted U" effect of physiological and psychological arousal on performance ability. With training, superhero Tony Stark is able to perform better and under more stressful environments than supervillain Obidiah Stane.

arousal). I have contrasted this "superhero" performance with that of a "normal" supervillain in the form of Obidiah Stane. Recall Stane was shown in all his villainy as the culprit who took over Tony's company, created Iron Monger, and then tried repeatedly to kill Tony. Not nice. Anyway, Tony's optimal mix of arousal leading to his maximum performance is shown by the arrow. If he moves beyond this optimal arousal level, his performance will decline. Obidiah Stane (see dashed arrow line) is shown as having both a lower maximum performance and a lower ability to function at high stress levels.

However, despite all manner of training and natural ability at managing stress and arousal, adding more stress to further increase arousal eventually leads to a decrease in performance. When attentional resources are at maximum, there can be increased errors. And piloting a hugely powerful armored exoskeleton loaded with weapons can be extremely dangerous. Let's talk next about the extreme military adaptation of Iron Man encompassed in War Machine, the name given to the most highly developed military armor that Tony invented. This armor debuted in "Legacy of Iron" (Invincible Iron Man #284, 1992).

The gist of the story line is that Tony Stark is apparently dead and leaves a taped message for his trusted friend Jim Rhodes to take over the company and assume the mantle of Iron Man. But Rhodey would be using a new silvery gray armor—the so-called War Machine armor. If you want to get technical, the original armor for War Machine was called "Variable Threat Response Battle Suit, Mark I" and the version for Rhodey was "Mark II Model JRXL-1000." I think it is easier to just call it War Machine armor for now. In an exchange that will have a lot of resonance for us later in this book, Tony explicitly explains to Rhodey that he "designed this last suit of armor specifically for you—to work with your own individual attributes, rather than mine." Even at this stage of Iron Man, there was some concession about how the armor needs "tailoring" for each user. (However,

A   B

Figure 4.3. War Machine armor as shown on action figures representing a character in the Marvel Studios film *Iron Man 2* (2010). Note the extreme militarization of the suit with shoulder and gauntlet cannons. Having this type of equipment built into the suit would make it difficult for the brain-machine interface to occur.

this is not addressed like this in *Iron Man 2*. In that movie, Rhodey just jumps in the suit, slugs it out with Tony, and eventually flies away!) We will come back to this concept. The War Machine armor is basically similar to the "standard" (hard to really use that word considering how many versions of the armor have appeared over the years) armor but with a real nod toward armaments. Figure 4.3 shows an action figure of War Machine as portrayed in the 2010 Marvel Studios movie *Iron Man 2*. See the attached mini gun on the right shoulder and the double barrel cannons on each gauntlet. While it may seem like a small add-on to the suit, having these as integrated parts or attachments on the armor would create some problems with control later on. Will these weapons become part of Jim Rhodes's body in the same way a prosthetic limb can? Or a suit of iron itself? And, does that hard suit of iron mean that the man inside gets soft? We deal with that in the next few pages.

CHAPTER

FIVE

Softening Up
a Superhero

WHY THE MAN WITH A SUIT OF IRON COULD GET A JELLY BELLY

I'll just stand if you don't mind. I've got to do something about the flexibility of this armor when I get back to my lab . . . [Later, crushes a cigarette offered by Jim Rhodes (maybe he knew smoking was bad for you?)]—oops! S-sorry about that. I guess I haven't got as much control over these gloves as I'd like.

—Tony as Iron Man describing problems with using the suit prototype, "Apocalypse Then" (Iron Man #144, 1980)

I have full mental control over the extremis armor—all the time. Even when it's deactivated. The trick is to zero in on the control systems . . . One part engineering, one part inspiration.

—Tony Stark, in "With Iron Hands Part 3 of 4" (Iron Man: Director of S.H.I.E.L.D. #31, 2007)

No doubt about it, Iron Man's armor is really cool. As I read the Iron Man comics over the years, I was always captivated by that armor. Then, the first Marvel Studios *Iron Man* movie rolled out in 2008, ratcheting up the gadgets on the armor. I like to imagine what a blast it would be to drive. Or wear. Or ride. Or merge with? Er, you know what I mean. Before you start building such a suit for yourself, we ought to discuss a few problems that might be created by throwing something as complex and sturdy as an armored exoskeleton on top of your trusty human body. First, since you already have a skeleton inside your body, putting on an exoskeleton means you would have to deal with two skeletons. That would create some real challenges in getting around! All throughout your life, your body has been well calibrated for how big you are and how your body moves. Second, wearing a suit such as Tony's would affect your ability to walk properly for a while after you took it off. Third, let's also not forget how uncomfortable wearing it would really make you feel. Rather than soft pliable clothing, your skin would be covered with layers of metal.

Next consider how that suit would affect your body temperature. Scientists call this "thermoregulation," or the ability to adjust your body temperature within a normal range. Kind of like a room temperature setting for your internal organs. Well, all the systems that affect body temperature are well suited (all right, pun intended!) for you when you are not wearing a huge metal suit of armor. Putting one on and then doing an extreme amount of exercise is likely to feel quite warm. (If you recall, that was my daughters' objection to wearing the armor. It would be too hot!)

Some of the issues Iron Man has to consider are pretty similar to those for deep-sea divers, firefighters who wear heavy and very hot protective gear, and astronauts. We will look at all these potential effects of wearing the suit and more in this chapter.

It All Depends on Your Underarmor

Wearing unpowered, that is, regular armor, would be really fatiguing. And awkward to move around in. This point was brought home to me quite clearly while writing this book. I had just watched (again) the fantastic *Lord of the Rings* trilogy on DVD. This time I also watched the special features on the third disc in which the actors are interviewed in the "Weta Workshop." Karl Urban, who played Éomer, said that when he first put on the armor "I went to walk and I just about keeled over. I was just so unused to the sheer weight of the costume." He also said that "once I put it on I was loath to take it off because it was quite a process. . . . Going to the toilet was fun . . . trying to fit into those small "port-a-loos" when your shoulders are 10 feet wide." (Despite its importance, we aren't going to touch on how Tony Stark goes #1 or #2 when he is in the Iron Man suit.)

Let's think about this further by using an issue I introduced in *Becoming Batman*: the effects of stress and physiological adaptation on the body. The body attempts to adapt so that the stress has a minimal effect. In *Becoming Batman*, I wrote about the stresses needed to produce Batman, stresses over and above what we would normally experience in our daily lives. Here I want to talk about the opposite: what happens if you remove the stresses? That is, imagine what would happen if we had fewer physical stresses on our body than normal. The bottom line is that our body systems work in a very predictable way when it comes to adaptations to stress. This means that removing the stresses also leads to a reduction in the benefits that occurred when the stresses were present. If we do exercise training that leads to stresses on our muscles, they get stronger to compensate for the stress. The reverse scenario is maybe a bit harder to grasp.

Iron Man Puts His Feet Up

A way that the effects of decreased use have been studied is to use bed rest. Yes, this means literally resting in a bed—as did John Lennon and Yoko Ono—for many weeks at a time. Prolonged bed rest doesn't remove all the effects of gravity—which provide crucial stress cues for keeping many physiological systems working well—but it does lead to an overall reduced activity level. The idea of using bed rest in people to study the effects of decreased use of the body came from work in other animals, such as the rat. Overall, there are some clear effects. The neurons in the spinal cord that make muscles work behave differently and are more difficult to activate than in an active animal, the protein content of the muscles decreases, and overall force output declines. At the same time, the cardiovascular system also degrades, so the amount of muscle energy and the ability to move are both reduced. I have highlighted a bit about what this would mean in figure 5.1. The first column lists different systems (e.g., muscle) or specific items (e.g., reaction time) and then the effects of either being generally physically inactive or undergoing prolonged bed rest, spaceflight, or habitual use of Iron Man armor. The important thing to note is the effects of using Iron Man armor shown in the far right column. The overall effect is universally decreased func-

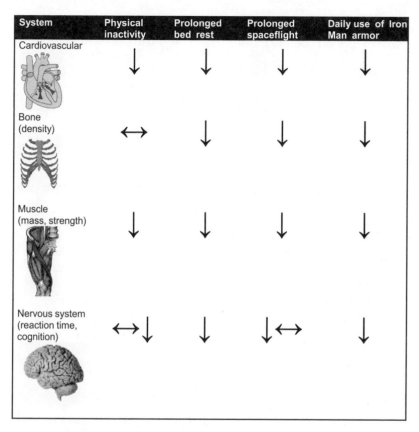

System	Physical inactivity	Prolonged bed rest	Prolonged spaceflight	Daily use of Iron Man armor
Cardiovascular	↓	↓	↓	↓
Bone (density)	↔	↓	↓	↓
Muscle (mass, strength)	↓	↓	↓	↓
Nervous system (reaction time, cognition)	↔↓	↓	↓↔	↓

Figure 5.1. The severe deconditioning effects of prolonged periods of wearing the Iron Man armor on Tony Stark and many of his body systems would be similar to those seen during bed rest, reduced physical activity, or prolonged spaceflight.

tion, with a startling result: using the Iron Man suit to enhance ability leads to a decreased functional ability of the human inside.

Out of This World

Probably the best, and most relevant, example to think about, though, is the literal "out of this world" experiences of astronauts

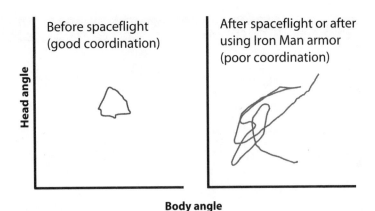

Figure 5.2. Using the Iron Man suit of armor for extended periods could disrupt normal bodily coordination. The left panel shows angle-angle diagrams for movement of the head and torso in astronauts before spaceflight. Note the tight and small area of the plot, which means very good coordination. The right panel shows the same concept plotted for astronauts (and implied for Iron Man) after prolonged spaceflight. The coordination is much weaker. Data redrawn from Paloski (2000).

during spaceflight and while working in space. Many of the stresses that help maintain the density of our bones and the strength of our muscles comes from the direct and indirect effects that the force of gravity has on us. When astronauts and cosmonauts are on long-duration spaceflights, they are working in an environment where the force of earth's gravity is essentially completely removed. This means that the stresses are removed, leading to weaker muscles and lower bone density. There are also problems in processing of sensation occurring during movement. The end results are overall weakness and reduced ability to coordinate the body, including the arms and legs. This can be seen in the plots of movement shown in figure 5.2, which shows how movement or position of the head is coordinated with the body (the trunk) before and after spaceflight. This tight coordination is seen by the close clumping together of all the points before spaceflight. In contrast, after spaceflight, there is much looser control between the different parts of the body. This can be seen by the much larger area covered by the plot in the diagram on the right.

Wearing the Iron Man suit for prolonged periods would also give rise to this lessening of control of how your body would move. This means that when Tony "doffs" the suit, he better not have to do anything that requires really good coordination right away. In the case of the Iron Man suit, the effect should be fairly short-lived and represents something called an "aftereffect." Remember the example of my running on the walkway in the airport and then the jarring experience of my landing on the hard tile surface from chapter 2? The effects of wearing the suit on Tony would be similar.

In addition, the effect shown in figure 5.2 gives an indication of how someone can respond to a perturbation. Most of your motor system responses, even if you don't pay attention to them, have to do with correcting body movements when there is an external perturbation. Think about riding on a subway train or a bus as an example. When the train moves, its motion causes a sway on your body. This is a perturbation. If you don't correct for the perturbation, you step or fall. The figure here shows how the chance of not correcting properly to a perturbation after spaceflight is greatly increased.

Enter—the Jelly Bellied Avenger?

Wearing the Iron Man suit of armor for long periods could be very much the same as the effects that would result from prolonged periods of being in space. This is where the "jelly belly" part of the title for this chapter comes from. Being in that suit would lead to extensive physical deconditioning. Especially in the later models, Tony is propelled into the air by turbines in his boots and external motors in his armor make him move. Since Tony Stark's body is in that suit, his arms and legs don't have to do much work anymore. This is why his experience is like being in space, where the reduced gravitational effects mean lower forces and less effort is needed for movement.

The force of gravity is seen as the acceleration of objects when falling (or being pulled). At sea level, this force is trying to accelerate objects at 9.81 meters per second squared, faster than many sports cars go from zero to 100 kph (60 mph). This may seem dramatic, but even while you are sitting reading this book, earth's gravitational field is trying to make your body (and the book, too, so please hold on) fall toward the center of the earth. Fortunately, this gravitational field is countered by the activity of your muscles. Your muscles act to maintain your posture and movements, and your bones are affected to maintain their mineral density and strength. The stresses induced by the gravitational field (and your movements within that field) that strain your muscles and bones help keep you the person you are.

Without that, you would become deconditioned. Your body must also work against the inertial aspects of your body. That is, different parts of your body have different masses, and any movement you make has to work against both the desire for that part of your body to stay at rest (or in motion) as well as the constant gravitational field. If you take a two-month visit to the International Space Station, you greatly remove these effects. As a result, you lose things like strength and bone density. This was recognized as a major issue for spaceflight and a large area of research into "exercise countermeasures" while in space has developed. Astronauts must do lots of exercise and working out in order to maintain their bodies as best as possible. It's basically a case of use it or lose it. But not losing it in space remains a major challenge.

If you exercise on earth, you are usually going to get in "better" condition. But if you exercise in space you are usually not going to get in "better" condition. Rather, you will at best maintain (but usually not really) or reduce the deconditioning effects. While sporting around in a robotic suit that moves his body for him—that is, the Invincible Iron Man—won't remove all the forces acting on Tony Stark's body, it will drastically reduce the impact of them. Extensive deconditioning of Tony's body can be expected, and he needs a rigorous exercise program to maintain himself. Or he really will get a jelly belly. With so much assistance to move, he actually would wind up doing less!

Rehabilitation Robots

Sadly, despite a childhood desire to be an astronaut, I was never able to experience being in space orbiting the earth. Sigh. However, I have walked around in a robotic pair of pants! This experience gave me an insight into how Tony might feel walking around in his suit and how disorienting it might feel to take it off again. In earlier chapters, we

saw that prosthetics help people who have lost limbs and neuroprosthetics help people when there is damage to the nervous system such
as after a stroke, blast-related head injury, or spinal cord injury. In
such cases, there are often many problems with being able to make
normal movements, such as using the arms in tasks like reaching and
using the legs during walking. Very recently neurorobotics has sprung
up as a specialized field in which powered exoskeletons for assisting
arm and leg movement are being used to help move the limbs with
external devices. Recall these are assistive devices, well, because they
assist with movement. If you have problems with movement, these
neurobiotic devices provide important help. But if you can already
move and have a robot amplifying your movements, it also does most
of the normal movement for you too!

Starting with the shrapnel in his heart in his origin story, Tony
Stark has had several health crises over the years that have resulted
in his needing assistive devices. One example is in a story arc spanning Invincible Iron Man #242–245. In "Master Blaster" (Iron Man
#242, 1989), Tony returns from a battle in which Iron Man once
again defeats his nemesis the Mandarin. However, now as Tony Stark
and wearing no armor, he is surprised to find his former girlfriend
Kathleen Dar has broken into his apartment. She pulls a gun and
shoots him in the chest in the last pages of that issue. We find out in
the next issue and the story "Heartbeaten" that Tony is not killed by
the gunshot wound but is left with a spinal cord injury. In the words
of his attending physician, we learn that "the bullet's passage destroyed vital nerve tissue along the spinal column . . . Damage that
even with today's technology is irreparable. As a result . . . Tony
Stark will never walk again!" (By the way, I have to interject here and
state that, in my opinion as a neuroscientist, all nerve tissue is vital.)

When it comes to walking and movements of the legs, a couple
of devices that you can buy—almost off the shelf—are the ReWalk
and the Lokomat. The ReWalk is one of the new kids on the block for
commercial robotic prosthetics. It is made by Argo Medical Technologies and is a programmable robot that can be set to produce
standing, stepping, and other basic movements. Lokomat is the name
of a robotic assistive device produced by Hocama in Zurich, Switzer-
land that is basically a set of exoskeletal "pants" worn by someone
and falls in the category for use in a kind of therapy called "body-
weight assisted treadmill training." Using a complex computer controller and a series of motors that can move the legs, the Lokomat

can produce stepping and walking patterns. The point of devices such as this in clinical use is to help people who have had a stroke or spinal cord injury retrain their walking pattern. After damage to the nervous system, overall muscle weakness is quite common. The Lokomat is used in walking to help support the body and then to move the limbs in patterns that are like walking.

I had the opportunity to try out the Lokomat for walking retraining by visiting the lab of a colleague, Tania Lam, at the University of British Columbia in Vancouver, Canada. Tania and I are both part of the International Collaboration on Repair Discoveries (ICORD) based in Vancouver. Research at ICORD is all about discovering cures and restoring functions for people with spinal cord injury. Tania let me walk on her treadmill set-up with the assistance of the robot. Despite knowing quite a bit about how it works, having seen the Lokomat in use many times, I had never actually climbed in and given it a try for myself. You can see an example of what this device looks like in action in figure 5.3. It can move the legs in a walking pattern but needs the person to be hoisted up in a body-weight support harness system (looks a bit like a parachute harness). The figure shows me getting strapped into the device that will move my hips and knees while I "walk."

Panel A shows me just getting the harness system on; in panel B you can see that the motor system (exoskeleton) is now strapped to my legs. In panels B and C I am just being lifted off the treadmill belt slightly (see my heels lifted), and in panel D I am actually being "walked" by the Lokomat robot. It can work as a kind of passive system, where I just relax and the Lokomat "steps my legs" for me, or it can assist my attempts to step. It can even be set to resist against my normal movements. It was odd when I tried to relax and let the robot step my legs for me. It was also very difficult to do.

Related to the use of Iron Man suit and aftereffects shown in figure 5.1, whenever I changed "modes" on the Lokomat—for example passive to active assist to resistance—it took a number of steps to adjust to the new condition and then several steps to get used to it again when we shifted to an older condition. I guess it would be similar to feeling what it was like to walk for the first time. Or like what it is like to walk in a new scenario, such as on an icy surface in the winter, a slippery wet area, or walking along a beach in strong surf. It takes a bit to adjust to (but you can do, of course) and then a bit to "unadjust" to.

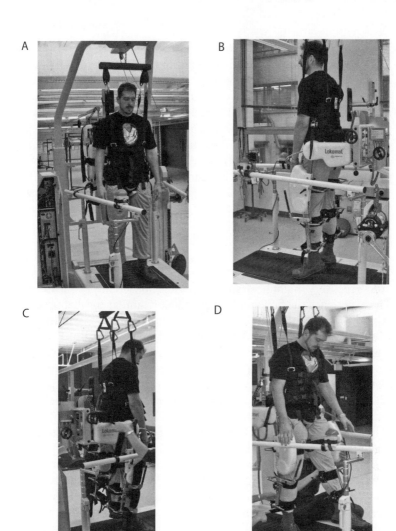

Figure 5.3. The Lokomat robotic exoskeleton. I am suspended over a treadmill and using the exoskeleton to move my legs. Things are just starting in panel A. Notice in panel B that my heels are off the ground as the harness system takes up some body weight. Panel C shows the apparatus from the back, and I am actually stepping—or being stepped by the robot—in panel D. This equipment can be used to help with walking retraining after stroke and spinal cord injury. Courtesy Tania Lam.

Back in Iron Man's world, Tony eventually reconfigures his armor so that, as described in "Yesterday . . . and Tomorrow" (Iron Man #244), "new servomotors and booster circuits move my legs for me! As long as I'm in this armor—I function as well as a normal man." In this way the Iron Man suit was used to restore function that was lost, not just amplify function that Tony had. A point he clearly reflected on in "The Doctor's Passion" (Iron Man #249) when he said, "Never thought there could be such pleasure in a simple phrase like, 'I'll walk.' But after the time I spent in a wheelchair when I was shot, just putting one foot in front of the other seems like a miracle!"

A "Neuro"-plastic Iron Man

Tony's recovery might be dramatic, but it is not actually miraculous. The experience the fictional Iron Man had is similar to what occurs in real life when the nervous system adapts to changes within the body. This is called "neural plasticity," and we will encounter it numerous times in this book. The big theme in this chapter is the use of assistive technology to amplify performance, specifically in the context of amplifying Tony Stark's abilities to produce Iron Man in action. But we are basing our discussion on examples of technology for real-life rehabilitation, such as devices like the Lokomat, which help retrain walking after spinal cord injury or stroke. This retraining of the body is directly related to neural plasticity.

Some time ago it was noticed that when a four-limbed mammal, like a cat, had a spinal cord injury that made it difficult to move, the back legs could be trained to walk again by stepping the legs on a treadmill. The animal then got better at walking. While the movement never becomes completely "normal," it can be functional walking. This shows the ability of the nervous system to adapt and change. It is quite different from the concept of the nervous system being "hard wired" and unchangeable. Instead, the nervous system should be thought of as highly adaptable and changeable. Kind of hard wired with soft wire I guess! Going back to the specific example of walking, as in other mammals the brain and spinal cord coordinate our arms and legs together. There is strong evolutionary conservation in these connections and in the basic circuits in the brain and spinal cord that drive things like walking. We have collections of neurons called central pattern generators (or CPGs) that are evolutionarily conserved across all species, spanning the swimming lamprey, the crustaceans, the cat, the nonhuman primates, all the way to our own species of *Homo sapiens*. CPGs are networks of neurons in the spinal cord that can generate simple walking patterns. We humans have flexible linking of CPGs responsible for each arm and leg. This is what gives us our ability to perform a variety of movements like walking, running, cycling, and swimming. In my own clinical neuroscience research, we look at how we can tap into these connections that can be broken or lost in people who have had strokes or spinal cord injuries. It seems likely that portions of the connections coordinating arms and legs can be still active after damage in stroke. This probably means that the remnants of these neural pathways can be strengthened with training. A lot of work shows that this kind of locomotor retraining can improve walking even many years or decades after injury.

What was immediately obvious when I attempted to use the Lokomat was that it was very hard to walk at first when the walking was being done by something else. This is an important thing to think about for this and the next chapter. Try to make a list of the times that you performed a difficult movement task like walking or reaching and you didn't actually have to do it for yourself. For most people such a list would be very short and might include no entries at all. My point is that usually when we want to do a movement we do it for ourselves. This means our brains are aware of both what we are trying to do and what actually happened. In contrast, when a robot moved my legs around during stepping, there was a mismatch between the sensory feedback that was occurring and my intentions for stepping, which were lacking. When these devices are used in rehabilitation of stepping in spinal cord injury, for example, it is likely that this disconnect is lessened. This is because there is already a separation in the relation between sensory feedback and motor output as a result of the injury.

This was shown clearly in the original origin story in Tales of Suspense #39 (1963) and revisited in "Why Must There Be an Iron Man?" (Iron Man #47, 1972). A panel from that story (figure 5.4) shows Tony talking about needing to learn how to walk again. This is a bit like the experience I had using the Lokomat. Except I didn't have a full Iron Man suit. Or any kind of weapons, dang it. Anyway, this panel shows part of that disruption in coordination seen in figure 5.2. In that figure, the main concern was how coordination

There would be mismatch between feedback from Tony's own body and expected feedback from the armor.

Tony adapting to using the original Iron Man gray armor. Tripping, falling, and poor movements would be part of the adaptation process. Extensive training would be needed to incorporate the suit into smooth and natural movement.

Figure 5.4. Tony Stark as Iron Man shown learning how to walk again after donning the Iron Man armor from "Why Must There Be an Iron Man?" (Invincible Iron Man #47, 1972). Copyright Marvel Comics.

would be disrupted *after* taking off the suit or returning from space-flight. However, coordination would also be disrupted when first putting the suit on (or walking initially with the Lokomat as I described above). In research studies this is sometimes known as walking in a "force field," that is, when the normal movement of the body is restricted or resisted against. This is shown clearly when Tony says "I'm like a baby—learning to walk all over again." Tony also says that "this armor's circuits are coordinated with my brain waves," which links back to our discussion of the Iron Man suit as a fancy neuroprosthetic brain-machine interface. But what are the long-term effects of using the Iron Man suit on Tony's nervous system? Above we talked about neural plasticity and recovery of walking using robotic exoskeletons after spinal cord injury. What happens if your nervous system is fine and you use a robotic exoskeleton anyway? What kind of plasticity occurs then?

As seen so far, the list of concerns for Tony is rather long and includes muscle atrophy, lowered bone mineral density, and loss of balance. Bottom line: he better not have too much strenuous work to do as Tony Stark unless he adheres to his training! Next let's peek into his nervous system and see how things are going in there. We will find out that we will soon have to add brain remapping and recalibration of spinal cord pathways to his list of concerns.

Brain Drain

WILL TONY'S GRAY MATTER GIVE WAY?

Hard . . . to move. Well, I expected that. This armor's circuits are coordinated with my brain waves just as any living human's brain controls his body. . . . I'm like a baby—learning to walk all over again.
—Tony Stark describing the original armor, "Why Must There Be an Iron Man?" (Invincible Iron Man #47, 1972)

Blast! I forgot that my command circuitry is more sensitive in this armor! I've activated a dozen systems with one thought! And what's worse, they're cancelling each other out! I'm not going—anywhere!
—Iron Man in space using prototype armor designed for extended periods outside earth's atmosphere, "Sky Die" (Iron Man #142, 1980)

$\mathcal{S}$tan Lee really likes the nickname "Shellhead" for Iron Man. What would it be like to "live" for a while within that real shell of iron? Here we take the next step in considering what would happen to Tony Stark's body while inside Iron Man's armor for long periods of time. Your body tends to adapt to things that happen to it. Even simple things like the sensations associated with wearing a new shirt or pants that may feel uncomfortable initially fade into the background with time. This is called "sensory adaptation." Well, what sorts of sensory adaptation does Iron Man need to worry about? Or does he need to worry at all?

Let's dig a little deeper into the brain. Earlier we talked about the arrangement—we called it a kind of "mapping"—of cells in the brain that control all the muscles of the body. Now, by map, I don't mean a literal map or a picture like the homunculus shown in chapter 3. Instead I am talking about the direct correspondence between parts of the central nervous system and the muscles and other parts of the body. Because those neurons are directly involved in the brain control of movement, we called that the "motor map." Producing effective commands for how and where we move is always informed by our senses, so it should be no real surprise that we also have sensory maps of the body. "Somato-" means "body" in Greek; these maps are called "somatosensory maps."

These maps are present from birth and are refined based on what we experience in life. And, what an experience it would be for the nervous system to interface with a robotic device like the Iron Man suit of armor. The cells that have this representation in the somatosensory maps are found in the part of the brain known as the—as usual, please wait a beat and insert dramatic pause for scientific creativity—the somatosensory cortex.

Here's a little thought experiment to help appreciate the extent of mapping of the brain cells in the somatosensory cortex. Imagine that you could record the activity of the neurons that are in that part of the brain. Now imagine that we are looking at the activity in the neurons of the brain receiving sensation from the skin of your left hand. Next, take your left index finger and tap it on the page that you are currently reading. The neurons of the cortex that are connected to the receptors in your left index finger that respond to touch would now be active. If you moved to a different finger, the same thing would happen to that brain area. You could continue to do this for all the skin areas on the body and you would create a (very distorted) map of the body. This is basically what the first scientists who did these kinds of experiments found. This mapping is now studied by using many different methods. One is to take many electrodes placed over the scalp in a form of electroencephalography, or EEG. Another is to record directly from the neurons themselves by putting electrodes directly into the brain or on the surface of the brain. Neurosurgeons distinguish areas that are meant to be operated on from those that aren't by electrically activating the part of the brain in question. Then, using the sensations or movements that come from stimulating the correct (or more importantly incorrect) parts of the brain, areas for surgery can be mapped. You could also get this information on brain activity from scans taken with functional magnetic resonance imaging (fMRI), a powerful technique that includes both anatomy and the physiology of the cellular activity.

With all this in mind (pun intended), we have a decent idea about how information from parts of the body make their way into the brain. And this information can be captured as a kind of mapping representation of the body. What we really want to talk about here is the extent to which these representations can change. It stands to reason that maps that are created as a result of our experiences may be changed when our activities and experiences change. This is bringing us right back to the idea of neural plasticity that we talked about before in the motor system. Now we are in the sensory system, which is essential to consider if we want to comprehend how integrating the Iron Man suit of armor with Tony Stark's brain could occur. It would not be enough for him to direct his armor to move with his brain. He would also have to be able to respond to stimuli in his environment.

An extreme example of Tony's brain-machine interface is in "War Machine: Weapon of S.H.I.E.L.D., Part 1" (Iron Man: Director of S.H.I.E.L.D. #33, 2007), when Tony jacks into an entire satellite! Tony Stark's trusty former driver and pilot Jim Rhodes is told, "Your armor hooks directly into the satellite. In effect, it'll be an extension of you. A part of you. But you'll be unaware of your body's immediate surroundings." The bit about how the satellite is is a "part of you" is something that comes up again later. Another example of what Tony Stark experiences with his direct neural interface armor is in the recent story lines found in "Invincible Iron Man" and penned by Matt Fraction. In "With Iron Hands, Part 3" (Iron Man: Director of S.H.I.E.L.D. #31, 2007), an interloper (Rahimov) who tries to use the Extremis armor is told by the computer that it is "compressing your cerebral cortex ... bringing your neurons closer together ... this will enable you to think more efficiently." Well, I can tell you this won't enable anyone to think more efficiently—your brain is already extraordinarily efficient.

Some Cranial Cartography: Malleable Maps in the Mind

To get the idea of how these somatosensory maps can change and how wearing and interfacing with a suit of armor might change the brain of a fictional Iron Man, let's look at some real-life but surreal examples: face transplants and phantom limb pain. The idea of how we normally make use of sensation from movement is nicely captured by Tony Stark's comments in "World's Most Wanted: Part 1, Shipbreaking" (Invincible Iron Man #8, 2009). When reflecting on controlling the suit, Tony muses "trying to operate this suit without Extremis is like trying to fly six stealth bombers at once." To connect with this, think about any of the times you have been temporarily disconnected from sensation in your nervous system—like when your leg or arm has "fallen asleep." Recall what it felt like and how difficult it was when you tried to move around or do something ordinary like walking and you get a pretty good idea of how odd this sensation really is.

Usually picturing Iron Man involves linking a biological body to a robotic machine. A real-life example that may help us is when we try to link a human body with bits from another body. Don't think Frankenstein but instead think reattaching an amputated finger or toe or transplantation of a limb or body part from a recently deceased donor. Or a partial facial transplant. In November 2005, a French woman named Isabelle Dinoire became the first person to receive a facial transplant. She had a horrible experience with her pet dog, who, during a long period while Isabelle lay unconscious on her floor, disfigured her by chewing and destroyed large portions of her face. She considered numerous options and decided to go with the pioneering approach of two surgeons, Bernard Devauchelle and Jean-Michel Dubernard of the Amiens and Lyon hospitals in France. After a grueling 15-hour surgery, Dinoire had a new face that had blood supply and sensation from her own body.

Beyond the staggering work involved in the procedure, the main reason to bring this up for our purposes is to imagine what this neural interfacing with a new face would feel like. According to a process known as "reinnervation," nerves in the periphery will grow to targets at a rate of about 1 millimeter per day. Within six months, Isabelle Dinoire began to have sensation in her face. Both sensory reinnervation, to the skin for example, and motor reinnervation, to the muscles of the face controlling movement of the lips and cheeks, would continue occurring over time. It is striking to note that even five years after the procedure the sensation from the skin of the transplanted areas does not feel the same as her own skin. John Follain of the Sunday *Times* reported that she described to her physicians that the skin from the donor is "softer. I'm the one who can feel it." This is despite the biological certainty that the skin itself isn't actually softer. It just feels softer now to her. This is likely an outcome of the reinnervation of this new skin by the axons in Dinoire's facial nerves. It also reflects the fact that even interfacing biologically similar but separate parts from another person into the body of someone cannot fully replace the normal intact connections. And instead can lead to odd sensations such as described by Dinoire.

Ghosts in the Iron Machine: Phantom Limbs and Phantom Pain Syndromes

Our somatosensory and motor brain maps have been reinforced and structured within us since we were in utero. Our nervous system has been carefully calibrated and tuned to us and our experiences throughout our lives, and it continues to work very well throughout our entire lifetimes. However, there are two dramatically different outcomes when the integrity of the maps can be challenged and plastic changes can occur. One is with decreased use, the other with increased use. And what represented the best and most powerful example of decreased use for the nervous system better than the complete amputation of a limb? A limb amputation is without doubt the most traumatic thing that can happen to the sensory and motor systems. On the sensory side, inputs that have been there since life began are suddenly gone. On the motor side, muscles that used to be easily activated by the nervous system no longer exist. This creates a huge mismatch between the sensory and motor systems as well as a mismatch between expectations and outcomes. It turns out that the nervous system isn't very good at forgetting about body parts that it used to have and this results in some strange things.

Let's begin first with the sensory maps we have. Think about the connection between neurons in the somatosensory cortex and receptors on your index finger. If there is a tragic and traumatic accident and a limb had to be amputated, there would no longer be an index finger with receptors to be connected to the nervous system and brain. However, the neurons in the somatosensory cortex still do exist and are expecting and awaiting information from the body. There can be an "expansion" of the maps such that areas that were nearby the now-disconnected brain regions take over and make use of the neurons in the old area. This means that there is remodeling of the cortical maps and is a useful response to accommodate the needs of the nervous system. Your nervous system doesn't like to have a gap in the map of the relationship between the body and the brain.

One odd—but not uncommon—effect of missing limbs is phantom limb and the related phantom pain syndrome. Sometimes the representation in the brain of the amputated limb does not fade away and instead gets taken over by some other areas and persists despite the limb amputation—enter the phantom limb. People suffering from phantom limb syndromes can have the distinct sensation that the limb exists and is there. They can feel itching and tingling and perceive weight in the limb, the exact opposite feeling of a leg that has fallen asleep. Phantom limb syndrome would be merely a nuisance if it were all there was to losing a limb. Unfortunately, what often goes along with it is a phantom pain syndrome. This means exactly what it sounds like: the person feels painful sensations that seem to be coming from a limb that no longer exists! As you might guess, this is a very troubling thing to experience.

How can you treat pain in a limb that doesn't physically exist? The work of Vilayanur Ramachandran and colleagues at the University of California at San Diego has taken an interesting approach to this—by tricking the brain into thinking the missing body part does exist. If a split mirror is set up (often a "mirror box" is used), someone sees the other side of their body on the mirror side. Figure 6.1 shows me using a mirror box at the lab of my friend Richard Carson at Queen's University Belfast in Northern Ireland. Notice that it appears from the reflections that I have two arms, but one is the reflection of the other. My other arm is hidden behind the mirror. This has been used with amputees with phantom limb syndromes, including phantom limb pain, to treat the symptoms. If participants carefully study their movements and do different tasks with the intact limb while looking in the mirror, they will perceive that the amputated limb is actually moving and feeling sensation. The best look at this is in panel C in the figure.

I can tell you that it really did appear to me that I was staring at my right arm even though it was only the reflection of my left. It is a

A

B

C

very powerful illusion. (Of course, in a real scenario, I would have been asked to remove my watch from my left wrist. It kind of takes away from the effect.) Using something like this, over time, the sensation of the phantom limb will often be reduced or disappear. This may seem pretty wild, but it appears to be grounded in the fact that of all our senses, our nervous system puts the most weight on vision. If vision informs us that something is happening, we almost always "believe" that over information from other systems. So, in this way the visual system can be used to trick the brain into realizing that there isn't any phantom limb to feel phantom pain.

Tony experienced something similar to this, described in "This Year's Model" (Invincible Iron Man #290, 1993). In this story, Tony Stark underwent a procedure to help "fix" his nervous system degeneration. One of the outcomes of this procedure—which involved a much closer link to his armor—was enhanced sensation. While lying in bed in the recovery room, Tony remarks that "there are additional . . . side effects. I'm experiencing alterations in perception. Everything is too sharp, too clear. Sights, sounds, textures are overwhelming." There is also the possibility of creating other weird outcomes associated with certain neurological syndromes such as "synaesthesia," which is also known as sensory substitution. This means you get a different sensation than you expect coming from some input. An extreme example would be if you could taste words instead of hearing them. In Part 6 of the 2007 Hypervelocity story arc, Tony described experiencing something similar, "a sensory backwash of synaesthesia hits one splinterself—as I taste and touch and smell unlocked datafiles." The closest story that hints at phantom pain (or probably more of a neuralgia—pain from inflamed nerves) was found in the *Ultimate Iron Man* 2006 graphic novel written by novelist Orson Scott Card. In this arc Tony is shown as having a skin condition that resulted in almost constant pain. Sadly, as part of the recurring theme of alcohol use and abuse in Tony Stark's life, he found

---

Figure 6.1. (*opposite*) The author using a "mirror box" apparatus shown from three different perspectives. Notice the appearance from the reflections creates an illusion that I basically have two arms but one is the reflection of the other. My other arm is hidden behind the mirror. This can be best appreciated from panel C. Courtesy Richard Carson.

that alcohol consumption reduced the painful sensations. But at con-siderable cost—as we will delve deeply into in the next chapter.

Is There Space in Shellhead's Brain to Store a Skin of Iron?

What would happen if we tried to dramatically increase the represen-tation of the body in the brain without subtracting something? We would need to know this to determine the feasibility of fully inte-grating a nervous system with an Iron Man suit of armor. We don't have any Iron Man acolytes to bring into the lab to study and get the definitive answer on this. Clearly we can predict that dramatic changes in how the nervous system works would result from connection to the neural interface with the Iron Man suit. The tantalizing question is whether the cost of this adaptation is possibly the ability to use the body. That is, will the connections for the suit be so strong that they replace the normal representation of the body? I think the answer has to be yes. If you take this far enough, it suggests that Tony's pro-tracted use of a suit of armor could lead to his inability to control his human body. Based on what we discussed above and how finely tuned our nervous systems are throughout our lives, Tony Stark's bodily control would be very robust. However, given the extreme nature of the kind of implant needed to connect with the nervous system, a direct brain-machine interface would certainly be the kind of—very unnatural—scenario in which this could be overridden.

Some recent work by Karunesh Ganguly and Jose Carmena at the University of California in Berkeley has addressed a related point. They were interested in getting at what sort of long-term changes in the brain might occur with the use of a prosthetic limb in a brain-computer interface. Recordings were made from neurons in the motor cortex of a monkey while it performed a task that involved reaching toward a target. Each session involved recording from up to a hun-dred neurons (upper motoneurons) from the motor cortex of each monkey. The researchers then watched what kind of activity oc-curred when the monkeys did a very simple reach.

To get a basic idea of what the monkeys did, imagine lifting your elbow straight out from the side while keeping your hand at the same height as your elbow. (Yes, lots of these kinds of experiments involve arm movements like this. Partly this is because they are easier to con-trol. Partly it is because the way the motor maps of the brain are oriented, it is easier to get at the cells for the upper limb.) Now place your hand out in front of your body and imagine straightening your arm out and flexing your arm to bring your hand back. You could try reaching to different places on a flat space right in front of your chest (a two-dimensional, side-to-side movement). These simple kinds of movement are the ones that the monkeys were trained to do by looking at a cursor that would light up to indicate to them what direction to move to next. They would then repeat these trials each day over a two- to three-hour training period for almost three weeks. During the reaches, the electrical activity of the neurons in the motor cortex was recorded.

From these recordings, it was possible to build a map of activity related to the direction that the arm moved. This makes sense, because the neurons were from areas of the brain controlling muscles of the arm. They should be active during reaching with the arm. In fact, the neurons are specially "tuned" (meaning they fire at higher rates) to certain movement directions, and this information can be detected from the brain and used to determine what the brain is trying to do with the arm. Normally, recordings like this would just give information about the output of the motor cortex. However, if you recorded that information with a specialized computer program, you could use it as a control signal for something else. Like a robot arm, for example. Or an Iron Man suit, eventually. At this stage, though, Ganguly and Carmena used this signal to control a computer display of cursors for where the monkey arm would be if it actually moved. This may start to get a little confusing.

The image at the top of figure 6.2 shows how the monkey moved its arm using the brain-controlled cursor. That means the cursor was moving based only on the commands from the brain—without arm movement itself. Think about this carefully—this means the monkey was controlling the computer cursor to move using a command that would normally come from doing the same movement with its arm. The monkeys were so well trained that they knew what to do to move their arms when a light appeared and generated the same brain activity to move the arm even though the arm itself didn't move anywhere. This is full brain control that is exactly reminiscent of what would be needed for an Iron Man interface to work directly into the nervous system.

In the top panel of figure 6.2 are data from Ganguly and Carmena's monkey study. Taking this idea toward the long-term effect

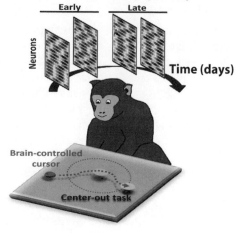

A   Stabilization of a cortial map with practice

Early   Late

Neurons

Time (days)

Brain-controlled cursor

Center-out task

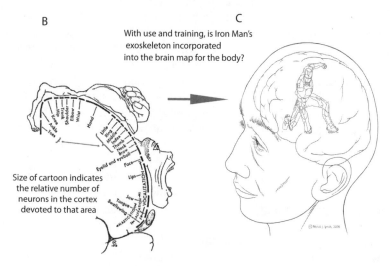

B

C

With use and training, is Iron Man's exoskeleton incorporated into the brain map for the body?

Size of cartoon indicates the relative number of neurons in the cortex devoted to that area

Figure 6.2. Continued use of a neuroprosthetic (like the Iron Man suit of armor) will lead to plastic changes in the cortical maps of the body. Changes in brain activity shift when a monkey learns a reaching task (A). The human motor body map, raising the question where would the Iron Man suit of armor go? (B). How the Iron Man suit of armor would have to be somehow incorporated into the normal body map (C). Panel A courtesy Sedwick (2009); panel B modified from Penfield and Rasmussen (1950); panel C courtesy Patrick J. Lynch.

of using the Iron Man suit of armor as a brain-machine interface is shown in panels B and C. The motor cortical map we saw in chapter 3 is redrawn here (panel B) so you can clearly see that integrating the exoskeletal suit of Iron Man armor into the brain means putting it on top of a normally "full" map! Panel C shows figuratively where the Iron Man suit might be represented in the brain.

It is an open question about how good could the brain representation of the neuroprosthetic actually become. Could it become strong enough and stable enough to become a real memory or an "engram" of a new map? This was the main meat of the Ganguly and Carmena study, and the answer seems to be yes. With practice over almost three weeks of training, the monkeys seemed to form stable cortical maps for the use of the prosthetic. It is surprising (to me at least!) that the neural activity seemed to be easily recalled and used, very stable, and very robust. These are the same features that would normally be described when discussing long-term memories. In fact, they use the term "prosthetic motor memory" to describe this outcome. Ganguly and Carmena go on to predict that with "improvements in technology neuroprosthetic devices could be controlled through effortless recall of such a motor memory in a manner that mimics the natural process of skill acquisition and motor control." These may well turn out to be prophetic words as we move technologically closer to a true Tony Stark / Iron Man neural interface. This could bring us face to face with the idea of embodiment of an entirely new body.

Can Tony Really Become One with Iron Man?

It is worth also touching on another strange outcome of wearing a suit of armor. That is again the idea of aftereffects. What I mean here by "aftereffects" is the continuation of a sensation or a perception after whatever you are doing to cause the change in perception has finished. This is directly related to the issue of the aftereffects of wearing a suit of armor that we discussed in chapter 5. So I mean effects that . . . continue after. Probably the best example of an aftereffect that you have likely experienced is a playground merry-go-round. When you spin around and around, you activate neural circuits related to balance information from the inner ear that carry on having effects well after you stop spinning. Recall trying to run forward after spinning on a merry-go-round. Although you sure do try to go forward in a straight line, you tend to deviate to one side despite your best efforts. Another example of similar vestibular aftereffects can be detected by being on an oceangoing boat all day with the sea rolling under you. If you do that and then sit down on a stable object like a chair at the end of the day, you often find that you can "feel" the roll of the boat even now that you are on land. This is where the idea of getting your "sea legs" comes from. Also, your nervous system is pretty specific about this. If you were sitting most of the time while at sea, the effects will be largest when you sit down later. What would it "feel like" to be Iron Man when the armor was off?

Let's go back to that concept of "embodiment" in relation to prosthetic limbs. The idea of embodiment is that the artificial bit—the prosthetic—becomes so fully integrated into the person and her perception of her body that there truly is no line dividing the two. In a clever experiment with an unusual outcome, a team of Swedish scientists headed up by Henrik Ehrsson recently provided an excellent example of this. They were able to create a stunning illusion in a group of upper limb amputees, which had been shown before in people without amputation and is called the "rubber hand illusion." They created a sensation of embodiment that a rubber hand was actually a real hand attached to the stump where the amputated limb used to be. The experiment was very simple (as are most clever and truly illuminating scientific studies) and basically involved hiding the stump from view while placing a rubber hand in view of the participants (figure 6.3). Next, the experimenters used small paintbrushes to rub simultaneously the index finger of the rubber hand and the stump of the amputee for two minutes. Later, just the rubber hand was rubbed—but sensation was felt in the hand!

Despite the fact that a similar illusion works well in able-bodied persons, the researchers were skeptical that it could work after amputation. They were therefore greatly surprised when amputees identified generally the same illusions. In fact, strong illusions were found in one-third of the participants. Interestingly, the illusions were more powerful the sooner after amputation that the tests were done. The most important point, though, is that this clearly shows both the tremendous plasticity of somatosensory maps and how they can be changed after damage. And, how vision can trump our other senses—like we talked about above for phantom limbs and phantom limb pain. By the way, they also examined anxiety in participants using skin responses and

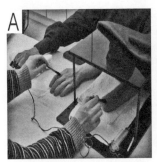

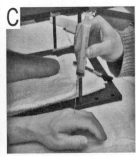

Figure 6.3. The "rubber hand" embodiment illusion. The stump of the amputee is stroked at the same time as a rubber hand that is in sight (A). Eventually the amputee perceives that the stroking of the rubber hand (with no activity on the stump) is actually coming from the stump (B). The illusion becomes so strong that in some people plunging a syringe into the rubber hand evokes physiological responses from the amputee (C). Photograph: Christina Ragnö, courtesy Ehrsson et al. (2008).

then plunged a syringe (panel C) into the rubber hand (which was not part of the body, remember). The participants had physiological responses of anxiety that would normally be present if the hand were part of their body. The strict relevance is that prosthetic limb designs that include sensors on the digits could be used to activate skin areas on the intact stump. Over time, this work suggests that the sensation from the artificial sensor on the prosthetic would become integrated into the perceptions of the person such that they are one with the body (enter, embodiment).

This raises the interesting idea that early versions (that is, before full nervous system integration) of an Iron Man suit of armor should have sensors on the fingers, hands, toes and whatever other body surface that activate skin areas on Tony Stark's body. In this way, Tony would really come to embody Iron Man in the way he declared in *Iron Man 2*, "the suit of Iron Man and I are one."

Incredibly, a recent study showed that including sensation from a robotic limb improved the ability to learn brain-machine interface commands. In 2010, Aaron Suminski, Nicholas Hatsopoulos, and colleagues at the University of Chicago used a "sleeve" over an animal's arm to help train monkeys to move a cursor on a computer screen based on recording activity in the motor cortex. This is just like the procedures for brain-machine interface we talked about back in chapter 3. The crucial difference was that the scientists at University of Chicago allowed the monkeys to use visual and somatosensory feedback together. Those monkeys learned how to control the cursor much faster and more accurately! I think this would likely have an effect on embodiment as well. This awaits future research.

Embodiment can arise from extensive use and practice with tools and devices. It is highly likely that this also occurs with extensive training with almost anything that is not part of the body naturally. This is also why learning to play a sport that uses tools—think golf or tennis, for example—is so challenging. This is probably because these tools or implements have not been extensively calibrated and mapped as parts of our physical bodies. Those somatosensory and motor maps have been continuously developed and recalibrated over all the years of your life to reflect your body size and habitual activities. But you use your body every day, and it has always been there for you. In contrast, the particular implements or tools that we use haven't been with us all the time and we don't use them continuously. Those maps we have for our bodies have to be able to integrate and incorporate the tools into our physical perception of ourselves. This is what is meant by embodiment. It does seem that, while these changes certainly do occur, the changes in our body maps are weaker than those for our actual body parts.

Because of that more practice—or maintenance activity—is needed to keep those maps strong and intact. In my own physical activity experiences in martial arts, I can certainly attest that complex techniques and movement patterns with empty hand are easier to initially learn and subsequently remember than are techniques and patterns based around weapons. As a result, it is much, much easier to lose track of, forget, or lose skill with weapons technique than it is with the empty-hand technique. Empty hand training uses all the "weapons" of the body that have been part of your body since birth. Weapons use involves tools. And what is a more complex tool than an Iron Man suit of armor?

By the way, this kind of embodiment and plasticity can occur even in actual tool use in humans with no neurological damage. Lucilla Cardinali, along with other French and Italian scientists, performed a fascinating study to look at this. They developed a long (about 40 centimeter, 15.5 inch) handled extension that had a "grabber" at the end. By squeezing the close end, it was possible to pick up objects and move them around. This was kind of like the grabber that can be seen everyday in parks and streets that cleaning staff move about picking up refuse and discarded items without having to bend down. In the study Cardinali and colleagues had people practice reaching with the grabber. Their research showed that using the grabber changed the movements of the arm even without using the grabber! Even more interestingly, the "aftereffects" due to using the grabber tool affected later simple pointing movements and also the perceived length of the arm. Participants had the impression that their arms were actually longer. The idea seems to be that the somatosensory schema of the body was changed by using the tool and the change had a general effect for many behaviors. Maybe even for Iron Man? In the 2007 *Extremis* graphic novel, after using the new suit Tony talks about how the suit is "wired directly into my brain. I control the Iron Man with thought. Like it was another limb." This would mean some process like that shown in figure 6.2 would actually have to be occurring in Tony's brain.

Iron Man in Space

Embodiment could apply to armored exoskeletons as well. Or spacesuits. I asked David Wolf from NASA about his experiences in space using a spacesuit. David is one of those astronauts who is an extravehicular specialist. That means he has spent a lot of hours literally in space, as in outside the spacecraft. His total time (including NASA and Russian MIR missions) is over 40 hours. In addition to that he has had more than 800 hours of water-based training wearing a spacesuit on earth. He has logged a total of 168 days, 8 hours, and 57 minutes in space. Approximately—but who's counting! Wolf explained to me that over time, wearing the suit is to "become one with the suit" (who does that sound like?). Eventually, wearing the suit "is like putting on your own armor and your old familiar tools. . . . It gets easier." This resonates very well with the idea of neural plasticity associated with learning how to use a new tool we talked about earlier. In this case a tool you wear over your entire body. By the way, this being a book anchored on Iron Man, I did ask Wolf about comic books and his favorite character. He likes Superman, for the record.

I also spoke with David Williams of the Canadian Space Agency. (All right, I admit it. Yes, I only spoke with astronauts named Dave in honor of the book *2001: A Space Odyssey* by Arthur C. Clarke.)

Williams, now a professor at McMaster University in Hamilton, Ontario (coincidentally, I am a MAC alumnus, just saying), told me that "you're as good as your training." It all comes down to the operator, in effect. He also emphasizes the role for mental imagery and rehearsal to memorize and perform sequences as rehearsed. This is to help "train like you fly, fly like you train."

Let's close this chapter with a few comments about how fun—or not—it would really be to wear a suit of armor for prolonged periods. Going back to David Wolf, he said that it "sure feels better" to get the suit off and you are "very happy to get out of the suit" and that "wearing it for long time is like being in a tight balloon." Additionally you have the concern about depending on the life support from the suit the entire time it is on. It is "probably like a race car driver when he gets out of the car after a race"; you are "hot, tired, hungry, and have been working at a very high level of peak concentration for a long time." In the comics this was also noted by Tony Stark's trusted assistant Pepper Potts. In the story "World's Most Wanted, Part 7: The Shape of the World These Days" (Invincible Iron Man #14, 2009), Pepper has made use of a custom-designed armor suit for herself (Tony is on the lam). She goes on to say "it's been nine hours in this suit, and, no offense, but if I don't get to get out and take a shower soon, I'm going to start screaming." With that, let's move on.

The Next Decades of Iron

"I Can Envision the Future"

The next phase of Iron Man project development would require getting around the biological delays and control issues related to relying on physical movement to drive Iron Man's activities. The steps in this phase each would involve a slightly more invasive integration with the nervous system. The first is to get stable motor control by using Tony's own nerve signals as triggers for activating the motion of the suit. To apply the existing technology of basic neuroprosthetics to Iron Man would require improved processing to more effectively use the nerve signals for controlling complex movements. This would take about three or four years. The next step would be to get around the delays and control problems that go along with using measurement of muscle activity. Tony would instead try to use direct brain commands for moving the suit and would need to develop a suitable brain-machine interface. The basic technology to do this exists, but it typically involves surface electrodes on the scalp of the human head which has limited control options.

These kinds of brain recordings provide only limited two-dimensional (e.g., up/down, side to side) movement control; they could only be used for controlling a paper-thin Iron Man moving across the pages of a comic book. To be more usefully applied in three dimensions, Tony would need to spend another five years improving the understanding of brain output as a signal for controlling the armor. He would then likely try to use the signal recording with the highest content: direct single-cell neuronal recordings. The technology to get basic information for controlling simple arm movements is available but has been mostly restricted to studies of the monkey. Some limited information has also come from studies of patients undergoing neurosurgery. Moving forward with this technology would take another five years. During this time, an improved knowledge of brain output from neuronal recordings would be needed, as would increased safety for recording from the human brain for prolonged periods. We are now almost 30 years into Iron Man development, and he would still be able to execute only rudimentary walking, lifting, and striking movements—and only by Tony paying strict attention and concentrating fully on the task at hand. More work remains. Tony continues to toil late into the night. Just about every night.

In the fourth decade of Iron Man R&D, we move into areas that exceed what is possible—at least currently, and maybe forever. There are some significant limitations that would arise even with an armor interface like the telepresence unit. To address these issues would require some kind of direct meshing between Tony's body (primarily his entire nervous system both sensory and motor) and the Iron Man armor, which could only be achieved through some as-yet unrealized nanotechnology. The closest description for this is the Extremis armor, and this technology does not yet exist. Not even a little bit . . . At all. The good news is that nanotech research, particularly biomedical applications, is a field that is rapidly expanding. An additional hiccup is how to deal with rejection by the immune system of any such interface and how the inflammatory response can be controlled so that the interface could be maintained over time. So, it is not possible to identify a timeline for this next stage. One last thing that Tony would need is the physical training required to become fully comfortable and competent with the Iron Man suit. This adds another five to seven years and gives a total projected timeline for inventing Iron Man and putting him into service of almost 40 years. Inventing, developing, and bringing Iron Man to reality would truly be a life's work. But what a life!

PART III ARMORED AVENGER IN ACTION

If we build it, what will come?

Trials and Tribulations of the Tin Man

WHAT HAPPENS WHEN THE HUMAN MACHINE BREAKS DOWN

> You know how dangerous a drunk is behind the wheel of a car? Imagine one piloting the world's most sophisticated battle armor.
> —Tony talking to Cap about responsibility while in the suit, "Civil War—Rubicon" (Iron Man / Captain America: Casualties of War #1, 2007)

> Ever notice how your PC can act funny and only you'll notice? An app takes a half-second longer to launch. It cycles longer on startup . . . I've started to notice little . . . things . . . with the suit. Not big enough to be glitches, and not big enough to trigger any alarm bells. But something's up.
> —Tony Stark reflecting on small problems with his suit, "The Five Nightmares, Part 1: Armageddon Days" (The Invincible Iron Man #1, 2008)

Being a superhero is hard work. It is even harder if you are a mere mortal and not an alien from the planet Krypton or the victim of an accident that leaves you with the ability to sling webs. Superman flies over the earth to decompress and Spidey climbs a tall building to get a new perspective. What does Tony do to cope with the demands of his chosen career? As you will see in this chapter, his solutions—frequently alcohol abuse—usually backfire on him. We will explore what happens when Iron Man gets buffeted by bomb blasts and when he needs to call the geek squad to fix bugs in his system. And I also have a little surprise for you in this chapter—a plausible origin story for Tony Stark's heart problem.

Not Shaken, Not Stirred: Intoxication and Iron Man Don't Mix

A drunk running around in an Iron Man suit. It is difficult to imagine something more poised for disaster than putting someone in an exoskeleton that amplifies strength and then having them get drunk and impair their ability—and their impulse control. This scenario played out in the Iron Man comic books and was shown in the extreme in the movie *Iron Man 2*. In both comics and in the movies, Tony Stark is always seen as having a penchant for a few drinks and even struggling with alcoholism. In the 2010 movie, Tony's drinking problems come to a head at his birthday party, where he drinks to excess and is intoxicated in the Iron Man suit. He uses many of the repulsor weapons systems and then gets into a huge battle with Jim Rhodes (which ends with Rhodey flying off in the War Machine armor). Let's consider the many ways in which alcohol would impair the performance of Iron Man.

Alcohol (in the form of ethanol) has some odd effects on the body. Initially and in reasonable quantities, alcohol generally can be relaxing and even somewhat euphoric. A large intake acts on the nervous system and produces problems of coordination, vision, and balance. It does this by affecting the brain (which we will look at more later) and the communication among the neurons themselves. Alcohol interferes with the coupling between excitatory (glutamate) and inhibitory (GABA) neurotransmitters and their receptors (NMDA and $GABA_a$) on neurons. This interference changes the way in which neurons function and respond to different inputs and varies depending on the dose (e.g., how much a person has drunk).

At lower blood concentrations, ethanol is known to affect certain types of $GABA_a$ receptors strongly. This effect causes much of the relaxation and suppression of behavioral inhibitions that alcohol initially produces. With higher alcohol intake, there is an interference with neurotransmitter binding to the NMDA receptor. Activity in this pathway is closely related to a cellular process called LTP (long-term potentiation) that affects memory formation and learning. This effect of alcohol is likely what gives rise to blackouts. When NMDA receptors are interfered with repeatedly as in chronic alcoholism, it may help stimulate a cellular response in neurons known as "apoptosis," which is a type of cell death. This helps account for some of the longer and more pervasive changes that happen due to the death, and thus the functional loss, of neurons.

It also seems that alcohol can affect the activity of sodium channels in axons. Since movement of sodium through this channel is vital to the excitability of a neuron and its ability to send signals down its axon, this is a very important effect. So, the story so far, too much alcohol is bad. Especially if it is chronic in the form of alcoholism. And, especially if you intend to hop into a robotic iron suit that connects to—and amplifies the ability of—your nervous system.

Beyond damage to the neurons, many other long-term physical changes to the nervous system can happen with chronic alcohol abuse. A common one is so-called nerve damage, which leads to "neuropathy." This basically means that axons in nerves are damaged and don't work as well anymore. Some of the more obvious effects of neuropathy are the slowing of information flow along sensory and motor nerves, which can result in sensations that are often felt (or not felt, actually) in the hands and feet. Also, even reflexes can be affected and this can make the control of movement much slower and of lower quality than ordinary. Chronic alcoholism can contribute to other significant changes in the central nervous system, including damage to parts of the brain essential for movement control.

While extensive damage can occur to many parts of the nervous system, I want to just focus on the cerebellum here. The cerebellum is vitally important in helping with regulating and altering movements that are ongoing and in helping with balance control. It is also necessary for correctly initiating and coordinating movement, particularly for the arms and legs. That means anything that interferes with or damages the cerebellum will impair the ability to produce movement. At this point, it is worth pointing out that more than one-half of the neurons in the brain are found in the cerebellum, this despite taking up only about 10% of the volume of the brain—those neurons are really packed in there! This means that the cerebellum is particularly sensitive to the effects of alcohol impairment.

In fact, the large suppressive effect that alcohol consumption has on the function of the nervous system is the underlying basis for the driving "spot check" tests that the police use if they suspect someone is driving while intoxicated. For example, a task like closing your eyes and reaching up to touch your nose requires the careful regulation given by the cerebellum. So does speech. As a result, if there is a problem with performing these smoothly, there is suspicion of alcohol impairment. Unfortunately, the flip side of this is that sometimes people who have had cerebellar disease or damage of the cerebellum are thought to be intoxicated when they are not.

Figure 7.1 shows a cross-section of the brain. The area of the cerebellum is shown traced for a normal brain (black line) and in one from a chronic alcoholic (gray line). This image is similar to one that would be obtained through magnetic resonance imaging. The cerebellum is the bit of the brain at the back and base of the skull and contained in the traces drawn on the images. Notice how much

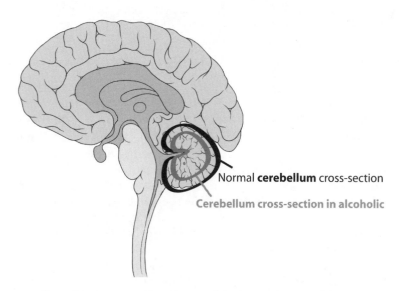

Normal **cerebellum** cross-section

Cerebellum cross-section in alcoholic

Figure 7.1. A cross-sectional image of the brain. The area of the cerebellum is shown traced for a normal brain (black line) and for a chronic alcoholic (gray line). Note how the cerebellum, which is involved with movement and balance, is much smaller in chronic alcoholics. Courtesy Patrick J. Lynch.

smaller the cerebellum is in the alcoholic. This clearly shows the anatomical changes that produce the behavioral deficits occurring in alcoholism. Alcohol impairment carries with it other effects on the nervous system, including cognitive problems that aren't present as part of cerebellar disorders, though. Some of the behavioral and cognitive effects that occur in alcoholics are thought to arise from other degenerative changes in the brain. This is particularly the case with links between the frontal lobes and other parts of the brain. The sum total of this is the impulsive behavior and cognitive problems seen in chronic alcoholism.

Let's take a minute to recap what we have been saying about the effects of alcohol on the nervous system and brain. When Tony drinks, the following tragic scenario is created: a man whose brain controls his powerful and weaponized armor is in danger of losing control. And of suffering brain damage. A man whose nerves must be "made of steel" to handle whatever life throws at him has nerve damage. A man who needs spilt-second timing as a pilot (of his own body!) and a crimefighter has impaired judgment.

Alcohol Addiction and the Iron Man

Throughout his whole comic book life as head of a multinational conglomerate, Tony Stark has often been portrayed as a devil-may-care, playboy socialite. In that way his character is very much like that of Batman's alter ego, Bruce Wayne. Except in the case of Bruce Wayne it is for show and for Tony Stark it is for real and true. While we always see Tony partaking in alcohol, for the most part he remains a "social drinker." Despite that, even when not totally in the grips of alcoholism, Tony still has many times used (flown) the Iron Man armor while drinking, which can't be good! In addition, there have been numerous notable periods where full-blown alcoholism threatened to destroy him.

The most (in)famous of these episodes was captured in a series of stories from 1979 written by David Michelinie and Bob Layton and with art by John Romita Jr. and Carmine Infantino. This story line was eventually taken to its bitter conclusion in the gifted hands of Dennis O'Neil (famous also for his work on DC Comics' Green Lantern / Green Arrow and Batman and Marvel's Amazing Spider-man and Daredevil among a host of other distinguished contributions) in the mid-1980s. The original story arc spanned Invincible Iron Man issues #121–128 and became known by the name of the last story "Demon in a Bottle" from November 1979. Facing problems with his company—Stark Industries, later Stark International—issues with the other members of the Avengers, his remote-control armor getting hacked, and girlfriend (Bethany Gabe) problems, Tony Stark spiraled into an increasingly alcohol-fueled lifestyle of a true alcoholic. As mentioned by Andy Mangels in his book *Iron Man: Beneath the Armor*, this story line was the first time a comic book superhero was portrayed having a chemical addiction. As an aside, this was a watershed moment for societal impact of comic book characters, and the Alcohol Information and Media Study Foundation praised the creative team for providing a story for youth about the proper use (and improper abuse) of alcohol. This is very much in keeping with the human element behind Iron Man in the form of the flawed—and authentically human—Tony Stark. Just as with Batman, another superhero with no supernatural ability, Michelinie is quoted as describing him as "simply a guy."

An interesting part of the theme of superhuman superheroes is well captured again by David Michelinie in the introduction to the 2008 hardcover graphic novel collection of the Demon in a Bottle story line. Michelinie writes that Tony Stark as a hero "will eventually do what he believes is right, no matter what the personal cost. And when he landed at the bottom of that bottle, and found the unbeatable demon that was his own weakness, he faced it, he fought it, and he drove it back into the darkest corner of his soul. And if there's a better definition of 'hero,' I don't know it."

It is, of course, this kind of connection with humanity that makes superhero stories so compelling both as entertainment and occasionally as parables. O'Neil's story "Deliverance" (Iron Man #182, 1984) portrayed Tony as having hit rock bottom and having been hospitalized. The cover for that issue dramatically phrased the seriousness of the problem with the text "in the morning Tony Stark will be sober or dead." It was a gritty and excellent story arc showing the harsh reality of alcohol addiction. Alcohol also figures in the *Ultimate Iron Man* graphic novel from 2006 written by novelist Orson Scott Card and drawn by Andy Kubert and Marc Bagley. Tony Stark is shown as having a bizarre abnormality where his nervous system—and not just his nerves—is found distributed across his whole body and heavily localized on the surface of his skin. The normal situation, of course, is that our skin is typically over the nerves. Here the nervous system was basically also on top of the skin. This facilitates control of the Iron Man suit (and is a perhaps inadvertent reference to the common embryological origin of neurons and skin cells) but creates even more neural tissue to be affected by alcohol. On the way to a gala, Tony is warned that "there's going to be a lot of alcohol there. Don't touch it. . . . Alcohol is a poison that attacks the brain. Your whole body has brain tissue. Everywhere."

It should seem pretty clear why some serious problems would loom eventually for Tony, given his abuse of alcohol. Alcoholism impairs the function of the nervous system, and the Iron Man suit needs to be connected to a nervous system that works well. At least if the Iron Man is to work well. And that is the whole point of this discussion. It won't work well. One of the things that research has shown is that even when the initial effects of alcohol on cognition and thinking are wearing away, there still remain problems of motor skill performance. That means that, even though someone (like Tony) may be recovering from an alcohol binge and be able to "think clearly," he would still make movement errors when doing something skilled (which would be anything and everything involving being Iron Man—or any other superhero, for that matter). This is a hugely important issue for performance as Iron Man and raises many concerns. Drinking and driving—whether ordinary vehicles on the road or a fully instrumented robotic suit of armor—do not mix. So far we have used alcohol as an example of a nervous system depressant. Let's look now at the opposite side of the coin—nervous system stimulants.

Stimulating the Nervous System of a Human Superhero

Just as alcohol is an obvious and widely used example of a nervous system depressant, caffeine is probably the best example of a stimulant regularly in use. Caffeine is used as a stimulant by many, often in the form of coffee or tea drinking. Caffeine has also been used as an aid to help improve physiological ability in sports, that is, as an "ergogenic aid." What is clear is that caffeine can increase wakefulness and alertness and can help increase performance in endurance exercise events. Caffeine also reduces the perception of how uncomfortable (or painful) or difficult a given exercise activity is. It may also help increase the ability of muscle to produce force during a contraction. There are some downsides to heavy caffeine use, however, including insomnia and stomach irritability. This is becoming even more of an issue in our society due to the increasing prevalence of caffeinated energy drinks. As Chad Reissig and colleagues at the Johns Hopkins University School of Medicine have pointed out, there is significant concern over the widespread and chronic use of high caffeine beverages. This is particularly the case since there is a strong relationship between high caffeine use and high alcohol use. Many companies have developed alcoholic drinks combined with caffeine. It's possible people who purchase these drinks believe that the caffeine as stimulant will offset the alcohol as a suppressant. So they then engage in things—like driving—that they might not do if they were consuming alcohol alone. But it doesn't really work that way. Jonathan Howland and his collaborators at Boston University compared the influence of beer, beer with caffeine, non-alcoholized beer, and non-alcoholized beer with caffeine on attention and reaction time in a driving simulator. Importantly, the addition of caffeine did not offset the reductions of reaction time and attention found with drinking beer.

Of particular relevance to our discussion of Iron Man is the military aspect of caffeine ingestion and how it affects performance. According to J. Lynn Caldwell and colleagues of the Air Force Research Laboratory, Human Effectiveness Directorate at Wright-Patterson Air Force Base, caffeine has been recognized in military aviation as a "stimulant/wake promoter" and is often used in the form of caffeine gum. Other central nervous system stimulants have been used by the military. In fact, all branches of the U.S. military services endorse the occasional use of dextroamphetamines and methamphetamines. Just so there is no confusion about what these are for, they are often called "go pills." However, higher doses of these drugs are only sanctioned in situations where the mission would be compromised and in the case of heavy fatigue. Chronic use is not sanctioned due to concern over health risk and due to blunting of the stimulating effect.

A horrific example of the use of stimulants comes from a military tragedy. This is relevant to our Iron Man example of War Machine in order to illustrate how vigilant Rhodey would really need to be. And how terrible the outcome could be if he wasn't. A tragic "friendly fire" accident occurred during the early days of the NATO mission in Afghanistan. This accident has been called the "Tarnak farm incident" based on the location near Kandahar, Afghanistan, and took place on the night of April 2, 2002. On this night members of the Third Battalion of Princess Patricia's Canadian Light Infantry were conducting a night firing exercise using anti-tank and machine guns at a firing range previously held by the Taliban. During the course of this firing exercise, an American F-16 fighter piloted by the U.S. Air National Guard was returning to base at the end of a ten-hour patrol. They mistakenly believed they were under fire from Taliban militants and took evasive action. They also requested permission from flight control to return fire. Flight control initially advised the F-16 crew to stand by and then subsequently to hold fire. However, despite that, a 227 kilogram (500 pound) laser guided bomb was released and hit the target on the ground. The result was four deaths and eight injuries to the Canadian unit, the worst friendly fire accident the Canadian Forces had suffered since the Korean War.

As a result of this tragic incident, two separate military boards of inquiry were launched, one Canadian, one American. In his book *Friendly Fire,* Michael Friscolanti details an important bit of testimony given by Colonel David C. Nichols. He is on record as stating that "combat aviation is not a science. It's an art." This is sobering testimony when we consider the tragic results that could occur if the Iron Man armor were misused or if the operator were ill disposed. Now, War Machine possesses armaments and destructive capability that easily would outstrip an F-16 fighter. An important outcome of both tribunals was that pilots were routinely given amphetamines (remember those "go pills," and yes, they also take "no-go pills" afterward) in order to maintain vigilance on long patrol missions. This was the case on this mission as well. Clearly it is worth considering how much the use of pharmaceuticals to maintain vigilance may impair the ability to make accurate human judgments under stress. Yes, a real F-16 fighter and a fictional Iron Man represent very impressive bits of technology. But they are both subject to the fragile nature of the real human beings piloting and guiding them.

Iron Man Suit for Protection

We have looked at all sorts of contraptions to help people walk, run, use their hands, and all sorts of prosthetic devices in our quest to determine whether it is possible to create an Iron Man suit of armor. I do want to touch on another aspect of the suit: whether anything exists currently that could offer the protection that Tony would need while battling bad guys.

Since Iron Man in the fictional world uses his suit as protection against bomb blasts and all manner of other kinds of explosions, probably the best example in real life is the suits worn for bomb disposal by police services and the military. In addition to the obvious concern for puncture wounds from explosions, an important consideration is the pressure wave associated with the blast. There has been an increase in terrorist and counterinsurgent bombings and improvised explosive devices (IEDs) in military zones such as in Iraq and Afghanistan, which has—unfortunately—provided many more observations on the multiple effects of explosive blasts on the human body. An immediate expansion of gas occurs when an explosion happens. This creates a shock (or "blast") wave that radiates out from the center of the explosion at supersonic speeds of about 3,000 to 8,000 meters (approximately 9,840 to 26,250 feet) per second. So, the blast wave actually hits things before any shrapnel or debris can get there. Four major classifications of injuries occur in a bomb blast: (1) those caused directly by the blast wave from a detonation, (2) "explosive" injuries from the shrapnel itself, (3) those occurring from the movement of the person or other objects at hand, and (4) injuries such as burns or radiation. Large-scale military explosives are designed specifically to enhance the blast wave. In an open space and with a smaller explosive like those commonly found in an IED, the blast wave effect may be smaller. In any case, an Iron Man suit of armor would need to be able to protect against all of these possibilities. A major issue is the pressurization to protect from the blast overpressure as shown in the top of figure 7.2. At the bottom of the figure is an example of the state-of-the-art bomb detonation units. In terms of how the bomb disposal suit looks, we are obviously a long way off from the cool Iron Man armor seen in the recent movies, but not that far off from the look of the original gray armor.

This pressure wave can cause severe damage to the throat, trachea, and lungs, as well as the ears and eardrums. Recently more attention has been given to the other body systems affected by blast waves. Amber Ritenour and Toney Baskin of the Brooke Army Medical Center, in Fort Sam Houston, Texas, have summarized that blast survivors may have eye injuries, rupture of the tympanic membrane (eardrum), lung rupture, intestinal damage, heart damage, and traumatic brain injury. It may be that the so-called shell shock syndrome

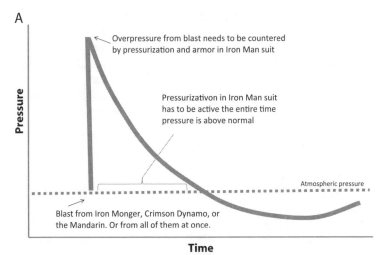

Overpressure from blast needs to be countered by pressurization and armor in Iron Man suit

Pressurizativon in Iron Man suit has to be active the entire time pressure is above normal

Atmospheric pressure

Blast from Iron Monger, Crimson Dynamo, or the Mandarin. Or from all of them at once.

Pressure

Time

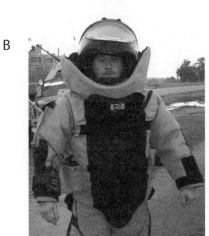

Figure 7.2. Pressure from a blast wave after a bomb detonation (*A*). Sample of the bulky protective suit worn by bomb disposal teams (*B*). Panel A modified from Ritenour and Baskin (2008); panel B courtesy the 31st Civil Engineer Squadron, U.S. Military.

is related to this. The effect of repeated exposure to blasts is something we will come back to again later on. The bottom line is that internal injuries are hugely important but have been typically underappreciated. An important thing is that the Iron Man suit would have to be pressurized and armored to such an extent that it could protect against the blast wave. In reality, this is a pretty tall order given everything we discussed above about what is currently available.

The Real-Life Iron Man Heart of Darkness

OK. I have always really liked Iron Man and his adventures in his own comic as well as part of the Avengers. In fact, Iron Man and the Avengers are among my very favorites. But, I have to admit I have never been very keen on the origin story with the chest plate that Tony has to protect his heart from the shards of shrapnel lodged in his pericardium or somewhere. Sadly, it just doesn't make much sense. (And, yes, I do appreciate the irony of picking out just one point to dwell on when we are talking superheroes here. But really, this needs thinking about!) In the spirit of talking about amplifying the body with technology that is actually available, I would like to propose a more plausible—but equally compelling and dramatic—real-life example. Instead of shrapnel near his heart, let's speculate that what Tony has instead is a life-threatening arrhythmia (literally meaning the heart rhythm is off), which means that the intrinsic electrical pacing that causes Tony's heart to contract and to pump blood is not working correctly. As a result of this arrhythmia, Tony's heart beats irregularly either too fast or too slow or with gaps.

The heart is composed of a special kind of muscle cell. There are also some cells called "pacemakers" that have the ability to help trigger muscle contraction without any kind of deliberate command to do so. A whole distributed network of these cells in different parts of the heart has the job of coordinating the contraction of the heart muscle so that blood is pumped effectively. When the first and smaller chambers (the atria) fill up with blood received from the body and the lungs, they contract and push the blood into the larger, stronger chambers called "ventricles." It is the contraction of the ventricles that pushes the blood from one side of the heart out to the body and from the other side into the lungs. Every time you feel or hear your heartbeat, an entire cycle of filling in the atria, priming the ventricles, and ejecting blood from the ventricles has taken place. This whole procedure has to be well coordinated to work properly.

The concept is similar to when you squeeze a tube of toothpaste before brushing your teeth. Imagine the atria are on the bottom of the tube and the ventricles are near the top. If you squeeze from the bottom of the tube, pressure will force the toothpaste out the top. However, if you squeeze in the middle, pressure will force toothpaste both down and up and much less will come out of the opening at the top. In your heart, you have valves to keep blood from going the wrong direction, but this example gets the basic idea across.

The most lethal of arrhythmias can arise when the ventricles or atria lose their pacing and "fibrillate." With atrial fibrillation, the heart muscle is essentially in a form of spasm and is not well coordinated. So, that normal flow from atria to ventricles is not very effective and blood pumping is compromised. This can lead to loss of consciousness and even death. We are going to assume that Tony has developed this kind of potential lethal arrhythmia and that he requires something to fix or regulate his heart rate. That something is an implantable cardioverter defibrillator, which we will call ICD for short. At this point you may be wondering what the difference is between a cardiac pacemaker and a defibrillator. A pacemaker helps trigger the heart muscle to contract (and therefore give heartbeats) by providing a rhythmic electrical signal. This can be set by a cardiologist and tuned appropriately for each person. Some pacemakers also have a defibrillator in them too. The job of the defibrillator is to detect when the heart rate has become irregular and then provide a strong electrical stimulus to reset the heart rate. So it works in just the same way as on the medical dramas where huge paddles are used on the chest wall, except much lower levels of electricity are needed since the implantable ones are right inside the body. I don't think Tony had need for a pacemaker but we are going to assume he needs an implantable defibrillator ready to step in and reset his heart rhythm if it changes dangerously.

Figure 7.3 shows what an implantable defibrillator looks like when inside someone. The wires from the implanted defibrillator (see label "electrodes in heart") snake down into the priming chambers, the atria, and into the main pumping chambers, the ventricles. While Tony is carrying on doing his daily work or sporting around as Iron Man, the ICD will continually measure and monitor the beating of his heart. When and if a series of irregular contractions is

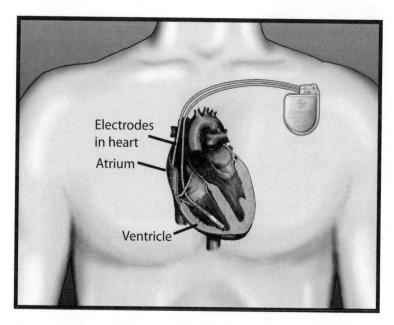

Figure 7.3. An implantable defibrillator. Courtesy Medtronic, Inc.

detected in his ventricles, the ICD will provide an electrical pulse to reset and restore the natural heart rhythm. The device also has the ability to increase the strength of the pulse if the restoration of heart rate doesn't occur. Although Tony and Professor Yinsen wired up the electric chest plate in Tales of Suspense #39 in 1963, it was actually in 1980 that the first human implantation of a defibrillator was performed. So, even if we revise the origin story a bit, it still presages actual scientific advances that came later! Iron Man science fiction and Iron Man science do link up.

What If Iron Man Goes Haywire?

Raise your hand if you have never had a computer problem. Ever. It could have been hardware or software—or maybe even neither if you were extremely lucky! If you didn't raise your hand, I am not going to say you are a liar, but maybe you have a selective memory? Almost all of us have had some form of computer malfunction from time to time. It has been frustrating and may have affected our ability to work or complete some project. Well, imagine what would happen if you had the same kind of glitches while inside Iron Man? Would Iron Man get viruses and what kind of malfunctions might he have? And what might it mean if such a machine ever ran amok?

This isn't a main theme for this book—it could be a whole book unto itself. But it needs to at least be considered here in the context of keeping things running smoothly. This very theme has been addressed numerous times in the Iron Man comics. A recent example was part of the humungous all-encompassing Marvel Secret Invasion story arc, which dealt with the Skrull having infiltrated earth's great superheroes and being poised to take over the planet. As part of the takeover invasion, they have managed to slip a virus into the Iron Man operating system. In the *Secret Invasion* (2008) graphic novel collection, Tony is shown inside the Iron Man suit—and not doing well. At all. The computer voiceover says, "Starktech armor: Complete system failure. Virus detected. All systems failing." Tony responds by saying, "I need to disconnect my bioware. The tech virus is hitting me like pneumonia. I have at least a hundred and two fever." This computer crash effectively takes Iron Man out of the picture.

An even more ominous example of a dangerous robotic suit of armor gone wild played out in Iron Man Hypervelocity. In this story arc, the idea of software and artificial intelligence is a major theme. Tony quips, "Ever get enraged by having to deal with poorly programmed and inadequately debugged software? Well, you should try being made out of software, like I am now." Later, the suit develops a mind of its own and goes berserk. Tony says "the latest iteration of my Iron Man armor has developed autonomous sentience and, presumably, gone rogue." This is a bit of a problem, as you could well imagine, both in Tony Stark's world and in our own. What kind of safeguards could be put in place to offset this?

Well, it is an open question but this issue was specifically addressed by the legendary and visionary author Isaac Asimov. In his amazing and influential 1942 short story "Runaround," Asimov laid down the Three Laws of Robotics: (1) a robot may not injure a human being or, through inaction, allow a human being to come to harm; (2) a robot must obey orders given to it by human beings, except where such orders would conflict with the First Law; (3) a robot must protect its own existence as long as such protection does not conflict with the First or Second Law.

Clearly the control systems for the Iron Man suit would need to have safeguards that incorporate these laws. However, Asimov did not envision (or maybe he did!) the kind of military application that Iron Man essentially presents. That is, Iron Man in action is routinely breaking the First Law! That is, causing harm or injury to a human being. Sure, they are bad guys, but how can this conflict be reconciled? Good question. And one for which I have no clear answer. Hopefully as we move down the road that could lead to such an item, someone else—maybe one of you reading right now—will come up with the answers. The huge problems of not having such a rock-solid safeguard were well shown in the 2010 *Iron Man 2* movie. Ivan Vanko, using Justin Hammer's technology, was able to hack into the War Machine operating system. With Jim Rhodes helplessly watching (or actually an unwilling participant inside War Machine), Vanko controls the suit to try and kill Iron Man—or Tony Stark really. This is big technology with big problems in need of big solutions. In the next chapter, we will look at whether Tony has the "right stuff" to come up with those solutions by exploring the mind of the "genius" inventor.

Visions of the Vitruvian Man

IS INVENTION REALLY ONLY ONE PART INSPIRATION?

I'll be even better . . . once you shoot me up . . .
with a reconfigured Extremis dose . . . I'm just a
man in an iron suit. I've spent months in my garage
trying to increase the armor's response time. And
it's still. Not. Fast. Enough. I need to wire the armor
directly into my brain. Extremis could do that. . . .
I need to be the suit . . . I need to grow new
connections.

—Tony talking to Maya Hansen, *Iron Man: Extremis* (2007)

Howard Stark: "It's big enough."
Tony Stark: "It's just a prototype, Dad. It's a lot
    easier to do the preliminary work when it's
    super-sized."
Howard: "What's the range? Flying, I mean."
Tony: "Fly! Give me a break, Dad, this thing weighs
    a ton. So far all it can do is hover and bounce."

—A conversation between father and son in a reinvented
Iron Man origin story, *Ultimate Iron Man*, vol. 1 (2006)

I have no particular talent, I am merely extremely inquisitive." So
said Albert Einstein (1879–1955), one of the most celebrated scientists

of all time. To this we can add the quote of Addison Altholz that "necessity is the mother of invention." The very birth of Iron Man in the hands of Tony Stark speaks directly to these themes.

Although Leonardo da Vinci (1452–1519) is most widely associated with his art (for example, that fairly well-known painting of someone known as "Mona Lisa" and that little sketch of "The Last Supper"), he has also been labeled one of the greatest, most innovative, and most forward-thinking inventors the world has ever seen. And with good reason. He combined many aspects of arts and sciences to become the original and prototypical Renaissance man. He created many machine-based inventions like the helicopter and the tank that just weren't possible to be implemented in his lifetime.

Sir Isaac Newton (1643–1727) represents another great example of genius, necessity, and invention. We usually see Newton being depicted as deep in thought under an apple tree. Then the apple falls, hits him on the head, and he is supposed to then be inspired to create a theory of gravity and gravitation. He is clearly one of the most influential—if not singularly the most important—scientist ever. What is not as well known is that, simultaneously with but independently from Gottfried Liebnitz (1646–1716), he also refined the mathematical language of calculus. Calculus was a necessity for the mathematics needed for Newton's studies of mechanics. So if we include invention to not represent solely a new device (Hey! I just invented a new mousetrap—here it is) but rather to include the development of fundamental and powerful new ideas, clearly da Vinci and Newton were great inventors.

Closely linked with discovery and invention is creativity. Have you ever spent much time thinking about creative thought? I am trying to get at that kind of thinking that—please suitably prepare yourself as I am about to use a hugely overused phrase—is often called "outside the box." Some people seem to have a real capacity for insight and creativity that they can tap into with tremendous success. In a book about the plausibility of different aspects of Iron Man, it is really worth asking how such an invention could be created in the first place. In my view, that means probing the basis for brilliance, discovery, and invention in science and engineering. Sadly there isn't really a recipe or user guide of steps to follow to become the next Benjamin Franklin. However, by examining a bit about the lives of some great inventors and pioneers related to Iron Man, I think we can take a glimpse into the process of "divine insight."

There are rather a large number of brilliant scientists and inventors in recorded history and it would be a bit much to talk about all of them here. So, for the rest of our exploration on invention and creativity, I think we should stick to those people who made discoveries that were fundamentally related to the main technology of Iron Man that we have been exploring: inventors who have done things that would help realize the vision of Iron Man.

Accidentally on Purpose

I must admit that, as a scientist, one of my pet peeves is the idea (mistaken in my view) that you can actually specifically target and deliberately direct creativity in science and engineering research and discovery. While we often partition out creativity in the arts as separate from creativity in the sciences, I think they represent one and the same deep process. This makes the opening of this chapter and the discussion of Leonardo da Vinci as an artist and scientific inventor highly relevant. What would da Vinci have produced if he was told "you have two years to make me a pretty picture"?

This also brings up a quote from a comedy sketch by my all-time favorite group, Monty Python's Flying Circus. In a sketch entitled "A Book at Bedtime" and aired on Monty Python's Flying Circus in 1971, we can hear: "Would Albert Einstein ever have hit upon the theory of relativity if he hadn't been clever? All these tremendous leaps forward have been taken in the dark. Would Rutherford ever have split the atom if he hadn't tried? Could Marconi have invented the radio if he hadn't by pure chance spent years working at the problem? Are these amazing breakthroughs ever achieved except by years and years of unremitting study? Of course not. What I said earlier about accidental discoveries must have been wrong."

This shows in some ways the opposite idea—that non-targeted discoveries get portrayed as targeted. I suggest instead that the process of creativity and discovery in science and engineering is a highly nonlinear and abstract one. It requires, to use that old expression, an open mind. A mind that is not constrained in any way, at least in the initial stages of forming the ideas. Implementing the ideas—putting the invention into play so to speak—clearly does have to be related to a realistic appreciation of what may be possible. In "With Iron Hands, Part 3" (Iron Man: Director of S.H.I.E.L.D. #31, 2007), Tony Stark is thinking about the direct neural integration of his armor when he says, "I have full mental control over the Extremis armor—all the time. Even when it's deactivated. The trick is to zero in on the control systems. . . . One part engineering, one part inspiration." The last bit is the most important. Engineering combined with inspiration. But what inspires?

And the Inventor Is . . .

It seems fairly clear that the skill set needed to invent Iron Man requires someone with an extraordinary and fertile mix. Creative thought, a strong will to succeed and pursue ideas at all costs, a technical knack for electrical and mechanical engineering, a deep appreciation of the neuroscience and kinesiology of how the body moves and works, and a ridiculous sum of money to enable the fulsome pursuit of all this. One of the interesting twists on the inventor angle for Iron Man is that Tony isn't just someone who creates the device but is also the very person who depends on the invention to survive.

I want to focus on three modern-day inventors and technological pioneers in this chapter. Of relevance to the idea of Iron Man and Tony Stark, these three have all developed and actually used the devices they created. They are Yoshiyuki Sankai, creator of the HAL robotic exoskeleton; Phil Nuytten, diving pioneer and inventor of the Newtsuit; and Yves Rossy, inventor and pilot of the jetpack fixed wing.

You will notice that I am limiting the discussion of invention and Iron Man to those inventors who intended to use their inventions and not just create something for others. Examples that come to mind that are relevant to the risky idea of robotic armor are the exploits of "birdmen," sky flyers, and the Wright brothers. With any development idea, the concept of trial and error testing is pretty commonly applied. This means slowly making incremental progress on a problem by testing whether some new change helped or hindered. What is less obvious possibly is the fact that this means the inventor risks being part of the error. And if the device you are working on is something that is implantable and links directly to your nervous system or allows you to make very large powerful movements—even flying were the technology and energetics available—that error could be fatal. I want to explore this idea to put in perspective just how spectacular an invention Iron Man would really be. And how extraordinary it would be for any inventor to actually survive the process right through to the end.

Since many of the images in Iron Man comic books and in the movies involve seeing Tony Stark flying around with his suit on, let's begin with the concept of human flight. Please don't think I have double-crossed you here! We did of course agree right up front in the introduction that the energetics and physics of really flying around in the Iron Man armor don't currently exist and are unlikely to exist any time soon. I still think we can use this as a metaphor for the creative process and dangers associated with inventing such a device. This brings us very close to the idea that genius and madness and derring-do of invention and foolhardiness are all intimately related. With that said, let's move on to the brilliant and daring Yves Rossy and his "fixed wing flyer."

The Skyflying "Jet-Man": Yves Rossy

Once upon a time a little Swiss boy named Yves Rossy (born August 27, 1958, in Neufchâtel) had a dream to fly like a bird. Lots of little boys (and girls, of course) have had a similar dream of being able to fly. (We can safely assume that Tony Stark had this dream sometime, perhaps many times, during the time after he developed the original Iron Man gray armor—although his initial dreams clearly involved also a whole lot of ammunition.) Often that dream goes unrealized and remains a really fanciful idea from youth. If it is realized, it often takes the form of becoming a pilot or in skydiving. Typically the dream would end there. And it did for a while for Yves Rossy. Yves was very fond of sports and did lots of running and biking but then became very interested in extreme sports.

He initially trained as an engineer and then qualified as a fighter pilot in the Swiss military. For eight years, he flew the fighter jets Hunter, Tiger F-5, and Mirage III. He then flew commercial jets like the DC-9 and Boeing 747 for 12 years. However, those more than 20 years of military and civilian piloting didn't relieve the itch of flying like a bird. Nor did his extensive experiences doing skydiving, hang gliding, paragliding, and aerial acrobatics. He needed to fly like a bird and that meant having wings and some form of power strapped to his back. So, Rossy became a skyflyer, someone who goes up in a plane as if to parachute to the ground but instead glides using wings.

As described by Michael Abrams in his book *Birdmen, Batmen, &*
Skyflyers, the ultimate goal is to be able to actually fly, not to fall
more slowly. That means covering lots of horizontal distance with
little vertical drop (think of a bird flying overhead—literally).

You can see Yves Rossy in flight in figure 8.1. Panel A shows Yves
on the ground, panel B is just after exiting the plane, and panel C is
under power and flying. Many have attempted to do this over many
centuries but one of the closest to this so far was Rossy. Using his
customized carbon-fiber-based wing design, in 2003 he covered 12
kilometers (7.5 miles) of horizontal distance while falling only 3 kilometers (1.9 miles). However skilled he is at "flying man," Rossy apparently preferred the idea of "jet-man" and so added a pair of kerosene-
fueled jet engines to his repertoire. Using the jets with his wings
allowed Rossy to make a groundbreaking flight on June 24, 2004.
While gliding down using his wings he then ignited his engines and
was able to fly horizontally (about 15 meters above the ground) at
almost 190 kmh (120 mph) for over four minutes. Yves Rossy has
now performed sustained human flight at speeds of over 200 kmh
(124 mph). In May 2011 he completed a historic eight-minute flight
over the Grand Canyon. These successes have not come without
risks, though, and that is something I want to highlight. Yves has
had his share of crashes. Since Yves is a pioneer in the use of jet-pack
flight with his fixed wing and this is the closest approximation we
actually have to Iron Man in flight, I thought it might be interesting
to talk to him. The following interview comes from that exchange.

EPZ: What interested you in creating a jet-powered fixed wing?
At what age did your interest start?

YR: Many years ago! Flying is a passion; it's a dream since I was a
young child. I always wanted to fly. I was a fighter pilot; I am
working now as a captain for Swiss Airlines, and I tried lots of
sports such as skydiving, for example. It was great but finally
it was only a fall. I decided then to create a wingsuit—to fly
longer—and then a wing, to fly level and finally to fly up! I
created a jet-propelled wing to realize my dream.

EPZ: How is the "feel" of the fixed wing compared to your
experiences in fighter jets?

YR: It is such a different experience—it is very hard to compare.
A 747, a Mirage Fighter Jet, and my wing do have a similarity
though. The decisive moments are at the same time, when you

Figure 8.1. "Jet-Man" Yves Rossy in action. Panel A courtesy Babylon-Freefly and panels B and C courtesy Blaise Chappuis.

take off. After that, it is only pleasure! The same in a 747, after the take off, you often fly using the autopilot. But I chose to devote all of my spare time to fly in the purest sense. Flying with my wing is an incredible feeling of liberty. It is awesome!

EPZ: What is the most dangerous moment you have experienced in inventing or piloting the fixed wing?

YR: The most dangerous moment is when I jump out of the plane and deploy my winglet (the foldable parts). I need a few seconds to stabilize my wing. It is the most difficult moment. I had a lot of incidents, but never a real accident. I have never been seriously injured. Because in case there is a problem, I can drop my wing and become a normal parachutist. That's why I never fly under 800 meters [about 2,600 feet], in order to have enough time to drop my wing and open my parachute. I always have a plan B in case of a problem, and if I am not self-confident, I don't fly. I don't want to take any unnecessary risks; I am cautious.

EPZ: Have you had any crashes?

YR: I had many failures! I had to drop my wing many times. But I learned a lot from my mistakes and from the bad test flights. Every incident allows me to optimize the wing. I have two parachutes. In case of a problem, I always have a plan. For example, I sometimes lose the control of my wing and there are oscillations. My military experience taught me how to move my arms in order to stop the oscillations. So, I'm ready to face each eventuality.

EPZ: Do you see the fixed wing as making an important contribution to society? That is, to push the limits of human ability or something else?

YR: I think it is an important progress, because from the very beginning every man has dreamed to fly. I also think that it is important to put the human back in the center, to refocus on the man, not only on the machine or robots. It's what I've tried to do with my wing, as did the pioneers in aviation, Leo Valentin or Clem Sohn, for example. The only flight instrument I have is a fuel lever and an altimeter, nothing else. I steer myself in the air thanks to the movements of my body: I turn my shoulders right to go right, and so on. Human species have a bigger adaptability than any machine. Machines must be the slaves of the man, and not the contrary!

EPZ: How much focus and attentional demand does it take and
could it ever be maneuverable enough to fly—like Iron
Man—as a fighter itself?

YR: I am more focused when I fly my wing, and I use all my
senses. For example, in a commercial airplane, you don't need
the sense of touch or smell. But with my wing, I need to hear
the noise of the engines, the hardness of the air on my skin,
and so on. In a commercial airplane, you need to be very
concentrated only for short periods: the take off, the landing,
if there is a storm, and so on. Then you usually fly with the
automatic pilot. Flying a fighter plane needs more concentration, because again there is a third dimension (you can fly
vertically, which you cannot do with a commercial aircraft).
It's the same with my wing; I can go right, left, down, and up.
The biggest differences are the number of senses you use and
the time you need to be very concentrated: from a commercial
aircraft (only a few senses and not a long time) to a fighter
plane and my wing (all the senses and all the time).

I am now developing a new wing, which will be smaller,
more powerful, and easier to handle. So, yes, it will be very
maneuverable, ready for acrobatics (looping and so on).

I know how to react in case of emergency. Not to panic,
but to think and react! So it's very intense, I am very concentrated, but not "stressed." I need to multitask, because I need
to to react and also to anticipate my trajectory. But to engage a
dog fight combat is not my aim at all. I don't want it. I've
already been contacted by some foreign armies, but I am not
interested in adapting my invention to a weapon. I would be
very happy to fly with other people, but only to share my
passion with them, to play and have fun in the air, not to fight
them! I am also thinking to take off from the ground, but this
will be for coming years. The only problem is the power; I
need to put on engines powerful enough to take off from the
ground and then fly. It's theoretically possible to take off from
the ground now, but then I will have no more fuel to fly!

EPZ: How much training would be needed to use the fixed wing?
Would you need full fighter jet training or could someone
without flight training being able to use it?

YR: You have to be a skydiver to be able to fly my wing. If you are
a good one, you can learn quickly (a few weeks, full time),

because you are already used to steering yourself in the air. I plan to teach people how to fly with my jet-propelled wing but only to experienced skydivers. I don't think that one day my wing will become a means of transportation. It will probably become an extreme sport, as the hang glider for example.

EPZ: As an inventor, what do you see as the key parts of your personality that have allowed you to go so far on this project?

YR: Discipline, perseverance, and above all, passion! I think it is important to try to achieve your dreams. To be able to go back and start from the very beginning when necessary. To keep believing in your projects and in your dream!

EPZ: Can you estimate the hours of development that the suit has taken so far?

YR: I worked 15 years on my project, building and testing more than ten different prototypes. I spent a lot of money and all my free time, but it really is worth it! Flying is a great feeling of freedom. The ten first years, I devoted all my free time working on my prototypes. But since 2007, I am on a sabbatical leave, so it's a full-time project! It really is impossible to count the number of hours.

EPZ: How many of the advances that were necessary to achieve the fixed wing occurred "accidentally" or by chance? That is, how difficult was it to plan the discoveries and technological uses?

YR: I work on my wing with this idea: "Learning by doing" (and sometimes crashing!). I always test my prototype, and when there is a failure, I try to fix it and keep working on it until it works. There were no advances that occurred by chance, nothing that happened accidentally in the development of my prototype. It was always the result of discussions, deductions, and tests. I learn a lot of every failure, so I can say that the advances occur thanks to the incidents in flight but not accidentally! I think that it is not difficult to plan the discoveries (it comes from the imagination); the most difficult is to make your invention become a reality and working!

EPZ: Lastly, Do you have a favorite comic book superhero? Do you know who Iron Man is?

YR: They are not superheroes, but I appreciate Mowgli and Baloo from *The Jungle Book*! For sure I know who Iron Man is. In the movie I recently saw, the problem to make it real is always the same—it's the power. We need to find a clean and power-

ful fuel, which allows me to fly longer with my jet-propelled wing for example. We have already the technology for becoming flying man or Iron Man. The main obstacle is to find an ecological or clean power—another power than oil. Technology without fuel or energy or power is nothing. We need to look for solutions and work more on the power.

Underwater Exosuits and Deep Diving Pioneer Phil Nuytten

Instead of protection from weapons and attacks, the kinds of exoskeletons now available have their origins in protection from harsh environments. Let's think about the harshest environment on earth—the deep ocean seabed—and the harshest environment not on earth—outer space. Some fascinating advances in hardshell deep sea diving and in spacesuit design are worth exploring. One of the first commercially and readily available exoskeletons was the "Newtsuit" invented by Phil Nuytten of Vancouver, British Columbia. Here we explore how deep-sea suits have gone from simple protection, which a passive suit can provide, to active enhancement of movement with powered segments. We also look at how astronauts use these suits in training in pools and in undersea diving.

Phil Nuytten has dedicated four decades of his life and founded multiple companies while improving and developing systems that maximize safety for use in undersea environments. He has striven to provide safe diving environments that have allowed access for scientific, military, and sport divers to get to the deepest depths of the oceans. His earliest work in the 1960s and 1970s was focused on the leading edge of diving physiology and on mixed gas diving. Nuytten began to explore the equipment and instrumentation end of diving quite vigorously in the 1970s. This included specialized life support diving gear for use in extreme conditions of polar diving. In 1979 his research led him to develop specialized diving suits for deep-sea application.

His first "Newtsuit" was a revolutionary hard diving suit that could be used at depths up to a thousand feet. It provides complete safety and protects from the crushing pressure at depth. It has been described as a "wearable submarine" and is thus clearly relevant to our discussion of Iron Man, the powered exoskeleton. In the late 1990s this was pushed to a just over 610 meters (2,000 feet) rated "Deep

Worker 2000" that was contracted by NASA for booster rocket recovery from shuttle missions. He continued his refining work to produce the "Exosuit" in 2000.

The Exosuit brings us very close to Iron Man in terms of flexibility and function. It is a "swimmable," ultra lightweight diving suit and a fantastic example of technical progression and invention. There are also plans to utilize a space version of the Exosuit, and astronauts from NASA and the Canadian Space Agency are currently being trained as pilots of the DeepWorker Submersibles. This project has implications for preparing astronauts for exploration on Mars or moon missions by training astronauts for work in extreme environments. In fact, Dave Williams, the astronaut we talked about earlier in chapter 6, trained extensively using the Exosuit and Newtsuit as part of his preparations for his Canadian Space Agency and NASA shuttle missions.

Related to the development of the suits, Nuytten's team has created a fully articulated, powered hand called the Prehensor. This articulated mechanical hand is meant for application on diving suits and pressurized space suits. Also, and this is shades of the "telepresence unit" Tony Stark developed, Nuytten's Prehensor can be adapted to remote-controlled interfaces. All of this work and more has strong applications for space exploration and military safety and clearly outlines him with skills as an inventor with relevance to Iron Man. Some of Nuytten's inventions can be seen in figure 8.2, along with Tony's version of an underwater suit (which he wore on top of his other armor).

Figure 8.2. (*opposite*) Deep water submersible diving suits, including the Exosuit, a swimmable exoskeletal suit including the Prehensor hand (*A*) and the Newtsuit exoskeleton (*B*) from the back (note the propellers for movement) and the front (note inventor Phil Nuytten getting ready to test his device in open water). Using a specialized exoskeletal suit (*C*) in "Deep Trouble" (Iron Man #218, 1987), Tony Stark commented that the conventional armor wasn't strong enough for application in the sea depths. Note that the Iron Man version of a deep water suit combines elements from the swimmable Exosuit and the Newtsuit. Panels A and B courtesy Phil Nuytten and panel C copyright Marvel Comics.

A

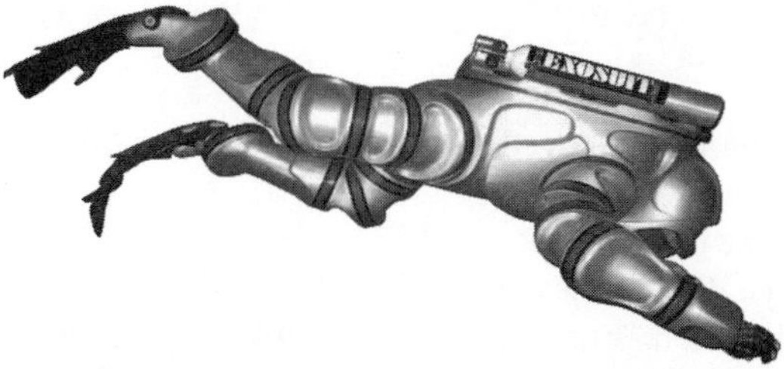
EXOSUIT

B

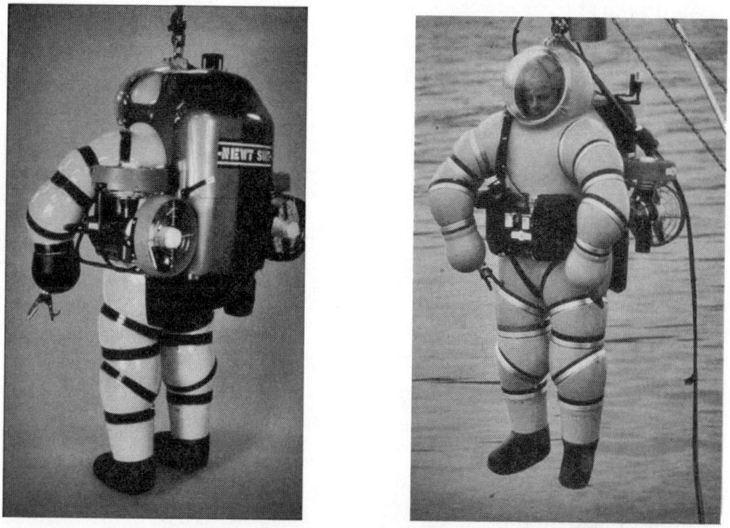

C

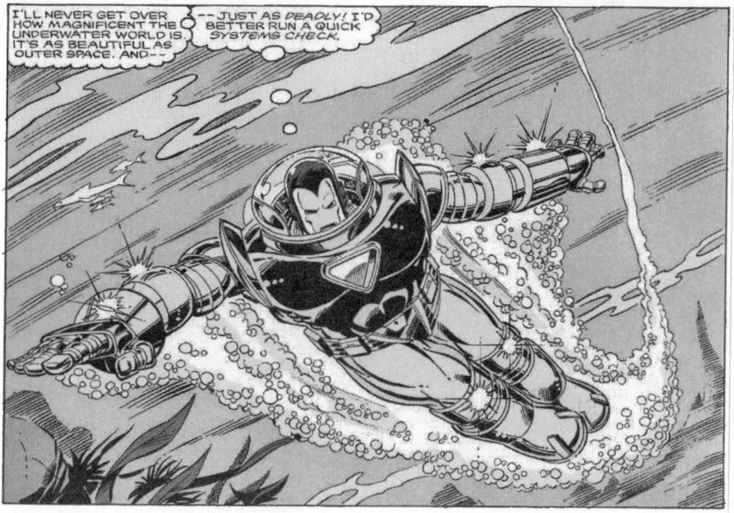
I'LL NEVER GET OVER HOW MAGNIFICENT THE UNDERWATER WORLD IS. IT'S AS BEAUTIFUL AS OUTER SPACE. AND--
--JUST AS DEADLY! I'D BETTER RUN A QUICK SYSTEMS CHECK.

Yoshiyuki Sankai and Belief in the Benefit of a Robot Suit

Based on the current state of the art (of the science!), I suggest (and break down during the course of the book) the whole process of creating and implementing an Iron Man suit—beginning with conceiving of the idea all the way through to development and training—could take at least 40 years. (We will look at what that means for the feasibility of Iron Man in the next chapter.) I have come up with these numbers by looking at related technical developments in neuroprosthetics and robotics. A good example to discuss a bit further is the HAL robot suit created by Cyberdyne Inc. in Japan that we have been talking about throughout the book. This suit has been the lifelong pursuit of Yoshiyuki Sankai at the University of Tsukuba in Japan. In 1968 as a young boy, Sankai read Isaac Asimov's book *I, Robot*. This idea of creating a useful robotic device that could be helpful to people captivated him and spurred his interest in electrical engineering. Sankai became even more fascinated in his elementary science classes by experimenting with frogs and how electrical stimulation could make their legs move (think back to our earlier discussion of Luigi Galvani and Alessandro Volta in chapter 3).

Throughout his youth, Sankai was fascinated by links between humans and machines and how such links could be used to improve human performance. Over time this became a firm commitment to help people with damage to the nervous system (such as after stroke or spinal cord injury) to become more mobile. He focused on the robot suit concept—not a robot because he rejected the concept of technological dominance implied by that name—and created a prototype in 1997 that could be used to help support the walking of a person inside of essentially what were robotic pants. The control for the motors was triggered by activity in the muscles of the legs during stepping, getting around many of the concerns about putting electrodes or other wires inside the human body. In 2010 the HAL (version 5) suit became available for use in limited applications in various fields such as physical rehabilitation and physical training support, for activities of daily living in people with disabilities or weakness, for assisting in heavy labor support at factories, and for possible rescue support. The current specifications on the HAL V5 suit are shown in table 8.1 and an image of the portion of the HAL suit for the legs is found in figure 8.3. To give some idea of how close (or, actually far away), we are to the Iron Man armor, table 8.1 shows

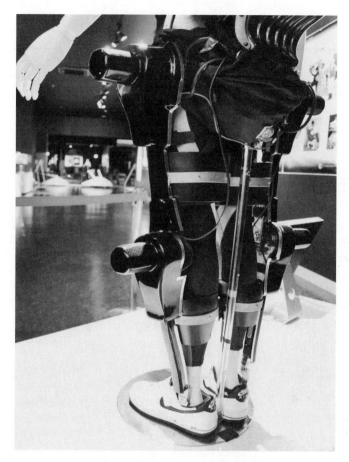

Figure 8.3. The lower limb portion of the hybrid assistive limb (HAL) suit by Cyberdyne Inc. Courtesy Yuichiro C. Katsumoto.

a comparison between the classic red and gold armor (see figure 1.2) and the Hal V5. By the way, it is interesting to notice how the HAL exoskeleton looks similar to the Lokomat we talked about earlier for walking rehabilitation. The huge difference is that the HAL suit multiplies strength and can actually be worn around all over the place. With the Lokomat, you can't really go anywhere!

I had an e-mail exchange with Professor Sankai and asked him about the development of HAL. He feels that his team has passed

TABLE 8.1. Comparison between the HAL robotic exoskeleton
and Iron Man armor

Specifications	HAL V5	Iron Man classic red and gold armor (Mark V)
Height	1.6 m (5′ 1″)	1.9 m (6′ 3″)
Weight	23 kg (approx 51 lb)	97.5 kg (215 lb)
Power	Battery driven AC 100 V	Battery; AC/DC
Continuous operating time	160 minutes	Many hours/days
Basic movements	Lifting and holding heavy objects; standing, walking and climbing	Basically anything including flying (technology does not currently exist)
Movement or strength amplification	Approximately 5–10 times	Approximately 75 times
Kind of operating control	Hybrid voluntary and robotic autonomous control	Automatic, computerized, "motion following"
Environments	Inside and outside	Everywhere including 20,000 m (70,000 ft) above and 300 m (1,000 ft) below sea level
Primary weapons system	None; no present or future military application	Repulsor rays (technology does not currently exist)

through many challenges in the development of HAL but they have now arrived at a version of HAL that represents the world's first cyborg-type robot exoskeleton that integrates the human body with a robot.

My discussion with Sankai echoed a fictional conversation in the Iron Man comics about the role for technology in the lives of humans. In the Invincible Iron Man graphic novel *Extremis* written by Warren Ellis and drawn by Adi Granov from 2007, Tony and Maya

Hansen go to speak with their former mentor Sal Kennedy. Sal asks what the point of their work is. Maya says "four years of engineering and I could cure cancer," while Tony answers that his time is spent thinking of "making a better Iron Man suit." This causes Sal to further comment "and a suit, Tony. Is that all it can be? She's working on military apps because that's how she's going to get the funding and the space to cure disease. What about you? What's the Iron Man *for*, Tony?" This question goes largely unanswered. Later, the Extremis origin story has a reboot with Ho Yinsen as a "medical futurist" whom Tony meets at a technology conference. He tells Yinsen that "the Iron Man program I floated at the conference is not about exo-skeletons or war. It's about becoming better. It's about bringing on the future. The earliest stages of adapting machine to man and making us great."

Sankai believes strongly that the combination of "engineering and medicine is the most meaningful when it helps human beings." Since he began his journey toward HAL he has wanted to supply leading-edge technology to people to support their normal activities, such as walking, standing up, sitting down, climbing up and down stairs, or doing heavy work. I also asked about what aspects of his personality have helped sustain him in the many years of work on this project. He said he really likes people, and he would like to develop technologies that make people happy. If you are trying to put together an Iron Man timeline, it is important to note that it has taken almost 20 years to go from the concept of HAL to a commercially available mechanized arm and leg robotic suit. The Cyberdyne project started in its infancy in 1991 and has continued to this day. Since there are and have been many developmental versions of HAL (similar to the many versions of Iron Man's armor), I also asked what the "end point" was. That is, what would the final version of HAL look like? The objective, he said, is to continue to "develop technologies that will help and make people smile. And we hope to create the future of the new field by developing a new HAL which nobody has ever seen before."

Tony Stark, Genius Inventor

According to Marvel Comics, Anthony Edward "Tony" Stark has a "genius" intellect. He has an advanced degree in electrical engineering. Those are his dry details, but what about what goes on inside the mind of an inventor? In the story "The Confession" collected in the 2007 graphic novel *Iron Man: Civil War,* Tony reflected that "I'm an inventor. I can envision the future. . . . I see what we will need and I invent the thing that will help us get there. That's how I invented my armor. That's how the Avengers were born. That's how every idea I've ever had in the world has come to be. I invent a solution."

How does all that creativity and inspiration work, though? Many people have reflected on this over the years. Dean Simonton came up with some fascinating thoughts on the subject in his book *Scientific Genius: A Psychology of Science.* Simonton examined many aspects of scientific discovery and the scientists behind the discoveries. Creativity is a key ingredient. So is the element of chance discovery. One of the interesting outcomes of this kind of analysis is that it is actually very difficult to specify in advance the discoveries that someone may want. That is, scientific discovery is a creative process and, like anything creative, cannot be fully scripted and demanded. This is sometimes underestimated in science, but it is considered commonplace in the arts. We routinely accept that novelists, painters, and actors need to find their "muse" or need to be inspired and then suddenly produce something genius. Science works in a similar fashion.

A key element of scientific advance and discovery is the role that random processes play. Related to this is the process of the oft-touted scientific method where faulty ideas are tested and rejected and far outnumber the rare observation of the correct idea occurring immediately in any field. Simonton has a great quote from William Jevons, who in 1877 said, "It would be an error to suppose that the great discoverer seizes at once upon the truth, or has any unerring method of divining it . . . the errors of the great mind exceed in number those of the less vigorous one. Fertility of imagination and abundance of guesses at truth are among the first requisites of discovery; but the erroneous guesses must be many times as numerous as those that prove well founded." Despite his success as the inventor of the first incandescent light bulb (and so many other inventions), Thomas Edison has been famously quoted on failure. When asked about the many thousands of failed prototypes he went through before arriving at a useful light bulb, Edison is quoted as saying, "I have not failed. I've found ten thousand ways that don't work." Part of this clearly relates to freedom of thinking and not being constrained by ideology. Albert Einstein did not talk much about the background process underlying his discoveries, but when he did he made it clear that the first stage of thinking was a very chaotic one in which many ideas that are not clearly linked together swirl around and take shape. Only after this are the ideas rigorously examined and tested.

How much formal education and what kind of grades would be required for someone to have the background to invent Iron Man? Well, possibly none at all! In his book on scientific genius, Simonton points out creativity is not necessarily well represented by high marks in school and college or university study. A distinction can sometimes be made between students who score well on tests of analytical power compared with those who are highly creative and intuitive. But it isn't a very simple and straightforward relation. Charles Darwin and Albert Einstein were marginal students in school but were shown to be brilliant scientists in practice. Contrary to that, though, Max Planck, Marie Curie, and Sigmund Freud were geniuses in their work and also had brilliant scholastic records. So, it depends. It also has been shown that with increasing levels of formal education there may be a reduction of creativity. This could be related to an increase in dogmatism that goes along with successful careers in science. However, it is possible that this constraint in viewpoint could be offset by other experiences in different fields and with different techniques and ways of thinking.

Philosopher Thomas Kuhn wrote about science, the scientific process, and particularly about the idea of paradigms in science, that is, the set of ways of thinking and theories in a field, in his classic book *The Structure of Scientific Revolutions*. A main point in Kuhn's work is that science moves along with a given paradigm until something very insightful is contributed by someone that leads to the breaking of the paradigm. Then a new paradigm is created. In this way, science continues to make incremental forward progress over time despite occasional setbacks. With regard to paradigm shifts, Kuhn wrote that "almost always the men who achieve these fundamental inventions of a new paradigm have been either very young or very new to the field whose paradigm they change." The upshot of this kind of thinking is that a scientist who has had training in different disciplines may be able to put ideas together in novel and unique ways that may dramatically alter a field.

Many preeminent scientists have had different interests that were harnessed very well in the pursuit of answering a given question. Nobel Prize–winning Spanish neuroscientist Santiago Ramon y Cajal (1852–1934) is an excellent example. Cajal pioneered the microscopic study of the brain and spinal cord and produced stunningly detailed and beautiful sketches and diagrams. But his primary passion was art and only because he was cajoled into pursuing a scientific career did he do so. Left to his own devices, he would have likely been a famous painter. He also had many other interests including boxing, fighting, and gymnastics. So, he was a very interesting person, indeed. He was the co-winner of the Nobel Prize for Physiology or Medicine in 1906, sharing it with another neuroscientist Camillo Golgi in recognition of their work on the structure of the nervous system. Throughout his life Cajal was a firm advocate of the "neuron doctrine," which describes the now well-known, but then controversial, idea that the nervous system is composed of individual nerve cells.

Another interesting point about creativity and discovery is that it does not seem to respond well to force. Simonton captured well the idea that solving a problem is often accomplished only after a long delay in directly addressing the problem itself. He quotes the famous French mathematician Henri Poincaré (1854–1912) commenting on his inability to solve a problem he was working on: "Disgusted with my failure, I went to spend a few days at the seaside, and thought of something else. One morning, walking on the bluff, the idea came to me with . . . suddenness and immediate certainty." Charles Darwin (1809–82), he of the theory of evolution, described a similar experience when thinking through a problem that suddenly resolved itself spontaneously, "I can remember the very spot in the road, whilst in my carriage, when to my joy the solution occurred to me."

In my own experience as a scientist, my most useful insights have occurred either when I had indeed been doing something else (like turning away in frustration from a grant or scientific manuscript and instead reading a Stephen King novel) or when I was engaged in doing something completely different—a change is as good as a rest. Given that I am writing this book on Iron Man right now, I want to share a neat thing that happened to me when I was writing *Becoming Batman*. Writing a general science book is a very different experience from detailed technical writing of scientific papers and grants. You are forced to take a very big picture view when doing this and while you do provide some details, you really try to make sure that you are taking an integrated view.

During the writing of *Becoming Batman* I was involved in some research activities and some reading different from my typical field of studies. I began doing some reading about the evolution of biped-alism in physical anthropology, which is outside my experiences as a neuroscientist but is directly related to my work linking arm and leg movements during human walking. Suddenly, or so it seemed, I gained insight into the evolutionary and mechanical link between rhythmic arm movement in human walking (which remains a key focus in my research program) and the role of the forelimbs of other primates during climbing. This allowed me to bring together many observa-tions I had not previously seen and kick-started another highly re-warding part of my research career. I am certain that I would not have made those connections at that time had I not been writing *Becom-ing Batman* and then been switching back and forth between differ-ent tasks. I firmly believe you cannot force creativity but that it must instead emerge over time.

Because of this I believe it would have been very difficult indeed for Tony Stark and Professor Yinsen to creatively design and create that very first Iron Man gray armor on demand in that cave in Af-ghanistan (movie) or Vietnam (comic book). Instead, I suggest that Tony must have been thinking for some time about this line of appli-cation. He must have had at least a vague inkling of a wearable, pow-ered suit of armor that he then brought fully to life with the help of Professor Yinsen.

Would the Repulsor Ray Be Repressed by Patent Law?

What if you could invent a working set of Iron Man armor? Would you want to make it and sell it? That is, when you weren't busy being a superhero. If someone comes along and actually does invent Iron Man armor, when will we next see it in Walmart? Nowadays the ca-sual inventor and professional scientists and engineers have to be versed in the legal aspects of discovery. And that means the concept of patents and patenting ideas and discoveries. An interesting point to consider is whether Anthony Stark would actually qualify as an inventor and whether the Iron Man suit could be patented. Keep in mind that Tony isn't just a superhero who actually tries to help people by acting as Iron Man. He is also head of a huge multinational tech-nological corporation. He needs, therefore, to help contribute to his company making money. And you make money by selling ideas as goods and those goods need to be exclusive to your company. This point was actually indirectly addressed in the Marvel Studios *Iron Man 2* film. It centered around how Jim Rhodes flew off with War Machine and gave it to the U.S. Air Force. A scene later in the movie, shows Pepper Potts (now president and CEO of Stark Enterprises) talking with someone at the air force base about getting back their "proprietary technology."

Let's begin with the definition of an inventor. I spoke with Cynthia Shippam-Brett, who is a patent agent working at a large firm in Vancouver. An inventor is defined as someone who contributes to performance of the "mental part" of the act of invention. This intellectual contribution goes beyond just conceiving of the desired result of the inventive process—in our case a fully integrated robotic suit of armor—but also an outline and understanding of the means necessary to actually accomplish it. So, the role of inventor goes well beyond simply coming up with the idea of "we should do this."

We could also think about who is not an inventor (in the case of Iron Man, this is a bit easier to consider because he mostly did it in secret). Someone who provided technical assistance, even if many hours were invested, or who supported morally or financially, or even someone who may have had an idea that stimulated the research itself. None of this qualifies as being an inventor. Tony Stark clearly is an inventor of Iron Man. Given the portrayals in the comic book origin tales and in the first Iron Man movie, Professor Yinsen would also legally be a coinventor of Iron Man. However, a bit of a legal battle could ensue in such a case because a key feature of inventorship is evidence to support the work, timing, and process. A lab book, carefully recording progress (and setbacks) and witnessed by others, is typically a key feature of this process. However, Tony didn't seem to document his work that well, so this could be difficult. This problem was neatly solved in most origin tales because Yinsen winds up getting killed in action during his escape.

Now, what about the next question—could Iron Man be patented? A patent is essentially a set of exclusive rights granted by the government that exists for a fixed period of time. The trade-off is that in order to be granted the rights, the inventor must publicly disclose the details of the device, the method of its operation, and what it is composed of. Patenting Iron Man would therefore mean giving a blueprint of the armor's composition and function. I think it should be clear that Tony Stark would never really seriously consider patenting Iron Man. Essentially it would mean providing for public scrutiny—

and in our Iron Man example this means to Hydra, the Mandarin, and Justin Hammer—the inner workings of the technology embodied in Iron Man. I don't think this is something Tony would be too keen on, really. Giving his mortal enemies a leg up on how to defeat him isn't really in keeping with a genius intellect! Of course all of this discussion presupposes Iron Man could exist in the first place. So why don't we tackle that very issue in the next chapter?

Deal or No Deal?

COULD IRON MAN EXIST?

> My god, the arrogance. . . . All those years refining a machine. Not a moment's thought given to refining the man inside it.
> —Tony reflecting on the use of the suit and the need for the human element, "Execute Program, Part 4" (The Invincible Iron Man #10, 2006)

> Undergoing the Extremis Procedure remade my body from the inside out. Long story short, my body was turned into a kind of computer designed to interface with the Iron Man. There was no longer a division between me and the suit. My brain . . . evolved, I guess. Into a kind of hard drive.
> —Tony Stark, "Godspeed" (Invincible Iron Man #9, 2009)

Iron Man belongs to that small club of superheroes that are viewed as "possible." This is a key part of his appeal. Iron Man and Batman are the two most obvious members of that group. A main attraction to Batman is that it seems like if you just worked really hard you could achieve his status (those of you who have read *Becoming Batman* will be already familiar with both the truth and falseness of this). A main attraction for Iron Man is that it seems like you could just pull on his suit and go flashing around as Iron Man. In some ways this is even more attractive for many than becoming Batman. At least those who prefer skipping the hard work of the training! So the main focus in this chapter has to do with exploring issues related to those perceptions. If the tech for a full-fledged Iron Man suit existed, could anyone become Iron Man? If you were Iron Man, would you need any training?

Could Anyone Just Grab the Garb and Go out as the Golden Avenger?

Answering the question of whether or not just anyone could actually be Iron Man is tricky. Throughout this book you have read my take on how and what might be needed for an Iron Man suit of armor to be controlled by a human operator. There aren't really any special limitations on who could be the operator, so long as that person doesn't mind undergoing the highly invasive interfacing required. So, the answer to that part of the question is clearly yes. However, can anybody just pull on the armor and rip around with it? The answer to that part of the question is equally clearly no. It would need to be integrated with that person and that would take some time. The more important consideration, though, would be learning to operate the system well enough to do something useful. Now, that would take a lot of training.

This discussion is relevant to some issues that have arisen numerous times in the Iron Man comics and now in the 2010 Marvel *Iron Man 2* movie in which War Machine figures prominently. One issue is whether someone could steal Tony's armor and then maraud around as him wreaking havoc or whatever. This has come up a few times in the comics but was addressed recently in a four-part story arc in 2009–10 called Iron Man Armor Wars. As a side note, writer Joe Caramagna and main artist Craig Rousseau do a neat job of retelling the origin story and doing nice homage to many of the early armors. This includes cleverly working in the original gray armor in "Down and Out in Beverly Hills" and showing a version of the original "Golden Avenger armor" (the one with the pointy face plates) in "How I Learned to Love the Bomb." It also includes an interesting twist with James Bond villain overtones with Dr. Doom lending a version of his armor to Tony just so that one day they can fight each other on even terms again. In any case, the main theme of the series is that some diabolical dude steals all the armor from Tony (including his new "Peace Keeper" armor designed only for search and rescue).

In this series, several bad guys climb into the armor without even warming up, let alone training in it, and then fight quite capably with it. The best example is Red Barbarian (Russian Madman), who uses the Hulkbuster armor with no problems at all! This is just not feasible given the amount of practice that would be needed.

Perhaps the example that most might be familiar with is the use of the War Machine armor in the comic books and in *Iron Man 2*. War Machine was created out of need by Tony to defeat the "Masters of Silence" who, in the words of Jim Rhodes from Invincible Iron Man #281 are "hi-tech samurai/ninja/kung-fu type hitters." As we talked about in chapter 4, the conventional armor wasn't, well, "armed" enough, so War Machine was created. This new militaristic armor was shown fully in action in the 1992 Invincible Iron Man #282 story called—wait for it!—"War Machine." Jim Rhodes's real debut in action as War Machine occurred two issues later in "Legacy of Iron" (Iron Man #284, 1992), in which Tony Stark is apparently dead (note: he is actually in cryogenic suspension). Tony's executor visits Jim Rhodes and makes him head of Stark Industries and lets him know that Tony has left him the War Machine armor for his use. Rhodey watches a video of Tony explaining what he has done.

Significant for what we are discussing here, Tony mentions that he is "asking you to take over." He mentions that "I've designed the last suit of armor specifically for you—to work with your own individual attributes rather than mine." So, the issue of how unique each set of armor would be is addressed at least partly, in the comics. This issue was also addressed in the 2009 20th Century Fox blockbuster *Avatar*. In the film, the identical twin of a scientist who was killed is recruited to take over his brother's use of a kind of brain-machine interface for remotely controlling a biological robot. It is interesting to see that the writers have worked into the script just how specific the neural connection for this kind of interface would need to be. Not just anyone could hook up and control the Avatar. Not just anyone can jump into the Iron Man suit and fly around either!

Jim Rhodes spent most of his career in the comics as Tony Stark's driver, pilot, and confidant. And, although he eventually became well known through his wearing the specialized "War Machine" armor (hinted at in *Iron Man* and shown very clearly in *Iron Man 2*), his first time in the tin can was actually shown in Invincible Iron Man #169 and #170 (1983) in two stories called "Blackout!" and "And Who Shall Clothe Himself in Iron?" The basic story is that Tony

Stark, who, as we saw in chapter 7, has a bit of a history with alcohol, has a few too many drinks, dons the Iron Man suit in order to defeat the villain Magma (who uses a huge kind of armored robotic spider as a weapon he can sit in). Somewhat impaired as he is, Iron Man is beaten up pretty good by Magma and returns to Stark Industries to regroup. This is where he meets up with Jim Rhodes. This is also the first issue where Tony Stark reveals to Rhodey that he is actually Iron Man (of course, Rhodey pretty much says he already knew that anyway). Then he has another drink (or two) and falls into a kind of fatigue and alcohol-induced stupor. Next, Rhodey really rises to the occasion. Since Tony is incapacitated, Rhodey, in true comic book dramatic fashion, takes it upon himself to put on the Iron Man suit and try to save the day. Also in true comic book fashion—and in the manner of a Shakespearean soliloquy—Rhodey comments aloud, "I just realized I'm putting on the Iron Man suit. Me. Jim Rhodes! Soon as I drop this helmet on, I'll actually be Iron Man!" Not quite, Rhodey. You will be a man in a suit of armor (look back at figure 1.1), for sure, but will you be able to function as Iron Man? In fact, in the very first panel of the next issue "And Who Shall Clothe Himself in Iron?" Rhodey says, "I've watched Tony Stark use this suit a million times, but I have no idea how it works." Too true.

Rhodey's reservations about trying a solo flight in the suit are a great lead in for our next bit here. The question I want to consider going forward is, would it work? How simple would it be to just pull on the Iron Man suit of armor and do something useful with it? In that story we just talked about, the first thing Rhodey does is to accidentally demolish a wall when simply trying to move his arm. Then he mentions that it is "weird . . . every slight twitch becomes a great big gesture—and if I even think about moving, I do!"

The bottom line is it is really very difficult to use the Iron Man suit, especially without the direct nervous system coupling that would be required. Recall that the main underlying point for how Iron Man could ever be controlled by a human being is along the lines of brain-machine/brain-computer interfaces. This allows our thinking to "piggyback" on many concepts currently being advanced. Another summary of this overall approach, and one that also shows the potential for use in rehabilitation, is shown in figure 9.1.

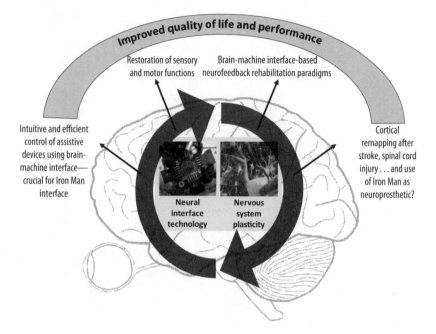

Figure 9.1. Summary illustration of the goals of brain-machine interface technology. These range from more simple applications to control computer cursors all the way to linking up a human to a full-fledged Iron Man interface. The field moves forward based on advances in biomedical engineering for neural interface technology and improved understanding of plasticity in the nervous system. Courtesy Doug Weber.

Real Life! Real Science! Boy, Iron Man Would Be Really Hard to Control

Let's take the best-case scenario and pretend that the potential user (following on from above, let's say Jim Rhodes) is already in good mental and physical condition and has lots of experience in many different activities. You might wonder why that is relevant. Well, the better physical and mental condition means the more effective and efficient will be any interface with a machine—like Iron Man. Also, more experiences means a larger skill repertoire to draw on that can be tapped into for using the Iron Man armor. So, assuming that, what is the big deal? Well, it is one thing to just interface with the Iron Man suit. It is another thing entirely to imagine being able to control that suit while it is running, jumping, flying, shooting weapons, performing evasive maneuvers, talking to others, responding to messages, regulating temperature, and so on. Even if we just take the most basic functions that Iron Man could possibly do, like standing up and walking across a room, and then reaching out, turning a door knob, and opening a door, we are talking about months and likely even years of training.

To come up with a "ballpark estimate" of the time needed for this, I spoke again with Jon Wolpaw. Recall that we met Jon when we discussed EEGs in chapter 3. Jon's work in brain-machine interface has largely involved brain electrical activity recorded from electrodes placed on the scalp. This interface allows for a good two-dimensional control system (think moving an image or cursor on a computer screen that has up/down and left/right as the two dimensions of movement). Such interfaces are on par with what has been achieved so far with electrodes implanted in the brain. I asked him about how long it might take a person to train and learn how to use this kind of mind-control interface. It seems that about eight to ten hours of practice time was the fastest to achieve a reasonable level of control. That training occurred in 24 sessions spread over eight weeks. Each of the three sessions each week totaled about 24 minutes of actual practice at moving the cursor in a targeted fashion. Jon advised that it is not uncommon for a person to take two or three times this time to achieve the same base level of control. They are currently studying ways to optimize the training but it is very difficult work.

Jon Wolpaw also spoke about something that I think is central to the issue of integrating a machine as intricate as the Iron Man armor with the complexities of the human brain and nervous system. And that is the issue not just of reaching some kind of level of control, or even of how fast it might take to achieve that control, but more of achieving consistent control. According to Jon, all of the methods, either invasive with electrodes right in the brain or noninvasive with electrodes on the scalp, have the same problem of "disconcerting unreliability. Control can be really good one day (or one three-minute period, or even one trial) and really bad the next." Jon goes on to explain that the "proper conception of BCI [brain-computer interface] use is not as mind-reading but rather as skill development [that is, it's like any other skill except it's executed by brain signals rather than by muscles]. The problem is that this skill doesn't become as consistent as muscle-based skills typically do. The fundamental reason for this is probably that BCI use is profoundly abnormal. The central nervous system evolved and is shaped throughout life to control muscles, period. A BCI asks it to control instead the signals (whether EEG or single neurons) from a particular cortical area. Thus, successful BCI use requires a major reorientation and a shift in the target of ongoing adaptive processes." These points are directly in line with what we discussed earlier about how unusual it is to superimpose another "limb" or device on top of an intact nervous system (think back to our discussion about phantom limbs in chapter 4). Wolpaw's work so far suggests that the human central nervous system "can do this (for some cortical areas), but not all that well and not very reliably." He suggests that the best way to "address this problem may be to use signals from multiple areas (thus imitating in some fashion how normal skills are executed)." This is a really interesting idea and gets at the concept of normal neuroplasticity in skill learning.

I also spoke with Doug Weber, another leading scientist in this field. He has experience with both invasive and noninvasive systems in both human and nonhuman primates. He told me that in the monkey it is possible to train "nearby 'patches' of motor cortex to control arbitrary movements." So, in this kind of experiment, monkeys could be trained to draw circles (remember this is drawing circles with brain signals, not actually using a pen or pencil!) over training with many repetitions and spanning numerous days. Doug suggests that a key issue that underlies the usefulness of machine interfaces is the quality of the brain signals that can be obtained. He said that it "is fairly easy to get good control in at least one dimension, when one or more electrodes are positioned in a brain area that exhibits strong, volitionally modulated signals—hand and face areas of motor cortex usually work very well, presumably because of the relatively large cortical map for those areas." Doug has been working with his colleagues doing studies in patients with epilepsy who have electrodes implanted in the brain for one to four weeks.

In epileptics for whom surgery is contemplated to remove parts of the brain that have abnormal activity, doctors often implant electrodes to try to monitor and locate the source of the abnormal activity. In any case, the patients also agreed to participate in additional experiments to help with brain-machine interface. In those experiments, they "start by finding electrodes over the motor cortex that show modulated activity during motor actions like hand open/close. Once we find a good channel, it is easy to couple the output to the control of a cursor movement. . . . The control is not perfect, but was achieved with only a few minutes of training. Performance improves with experience, but we have not had enough time to really study this in humans."

Weber offers a further reality check when he goes on to say that it is difficult to really predict how complex a device—think the complexity of Iron Man—could be controlled with any such approach. That is because all the studies to date have used very simple tasks and participants have solely focused on a single task. The point is that this is "pretty artificial, since we rarely focus so much attention on simple or even more complex motor tasks. No one has really studied how BMI [brain-machine interface] performance changes when the user is performing other tasks in parallel. 'Multi-tasking' would require at least a portion of the BMI control to be performed subconsciously." This last comment about multitasking is particularly relevant when we think back to our discussion of how attentionally demanding that would be in chapter 4.

The bottom line of all of this is that upon first "jacking in" to the Iron Man suit any user would probably be able to fairly quickly (within a few hours) learn how to do something straightforward like opening or closing one hand. How long to do something even more complex like standing up and walking? Using benchmarks from neural rehabilitation and attempts at relearning skills like walking after spinal cord injury or stroke, it is reasonable to estimate that this could be achieved in about three months of training. That might give enough time to provide the ability to stand up slowly and to walk at about one-half the normal pace across a space of about 30 feet. That is pretty sobering when we think of what is shown in the comic books and movies. So, running in and grabbing the suit, throwing it on, plugging it in, is just not possible. At least, not yet.

I can see why the writers and artists show that, though, instead of a panel that says "three months later" and shows the thief slowly shuffling across the floor in the stolen Iron Man suit! Not very thrilling, but that's the truth. Even the comic book writers did acknowledge a bit of this. Returning to the story "And Who Shall Clothe Himself in Iron?" (Iron Man #170) we began with above, Jim Rhodes does find he is having difficulty just moving around. So, he tracks down a scientist working in one of Tony's labs for help. All we see is that some time later, the scientist tells Rhodey, "like I said it'd take months—maybe years—to dope out all that circuitry." Umm. Yes. At the very least—for now. Just as the first "gray armor" Iron Man presaged things that are just appearing now, perhaps we will see a more intuitive control system that can be easily learned.

You might ask at this point why the suit couldn't just remain a device that only triggers its activity from the activity of the user. Kind of like an amplified passive suit like the ReWalk we first discussed back in chapter 2. This design could work for simple and very slow tasks like walking carefully across a room with a smooth floor. However, to do sophisticated things like maneuver out of the way of Whiplash or fight Iron Monger, it would be far too slow.

Your biological body makes use of many different timescales when you move. Your neurons work with a millisecond (thousandths of a second) scale, your muscles use tens of milliseconds, your movements take hundredths of a second, and your overall impression of what is happening occurs on a "seconds" timescale. You are fully calibrated for this in the same way that your brain body maps are calibrated to your body parts as we learned in chapter 3. In order for the Iron Man suit to respond in a way that you or Tony or Rhodey could control effectively, it would need to work on command signals for your neurons on that millisecond timescale. Anything further downstream (like using actual movements of the body to trigger the motors in the suit) would introduce huge delays. Using that approach would be like trying to move around and fight Whiplash with the kinds of feedback delays you used to get on a poor transatlantic telephone connection but which is nicely mimicked now by using a low bandwidth Internet connection to have delayed Skype conversation. Everything would be out of synch. Even small delays would be catastrophic.

What Happens If Ol' Shellhead Shorts Out?

Now let's move on and think about what kind of real-life examples of training would be needed. So, this means we have dealt with the fairly difficult problems outlined above and assume that the suit itself can be controlled—remember this is really just saying that the user has the ability to use his or her body. Now we have to superimpose all the technical training on top of that. This issue was hinted at in the 2008 *Iron Man* movie in a scene where Colonel Jim Rhodes is touring some recruits through a hangar and discussing the role of pilots and advanced fighter jet technology. Rhodey says, "In my experience no unmanned vehicle will ever trump a pilot's instinct, his insight—that ability to look into a situation beyond the obvious and discern the outcome. Or a pilot's judgment." Just after that, Tony Stark strolls in and says, "Colonel . . . why not a pilot without the plane?" Let's pause for a minute to think about how much training just that part would take.

To think this through, let's use qualifying as a jet fighter pilot as the example of training. Typically you need at least a bachelor's degree. You would have to be in top physical and psychological condition. OK. Assuming you got accepted and completed officer training school, the next step would be flight school. There you would study aerodynamics, aviation physiology, engine mechanics, principles of navigation, and land and sea survival. Then, during primary flight training, you spend many hours in the air (with four solo flights) and many more in flight simulator training. That would not include additional lecture hours for flight support instruction. Upon completion of this training, the top, top candidates would go on to five more weeks of jet training "ground school," which could include classes in meteorology, rules and safety of visual flight followed by actual flight training in aerobatics, communications, weapons use, and specialized takeoffs and landings. Next is advanced flight training, in which jet fighter pilots perform combat maneuvers and learn about flying at night. You get your wings only after five to seven years of training. That's a lot of work. But every step here is a step that an Iron Man "trainee" would absolutely need to complete.

Iron Man is often shown having kind of standard "mano a mano" fights as well. This was clearly shown in extended sequences of hitting, blocking, punching and kicking between Iron Man and War Machine in *Iron Man 2*. He would need at least some basic training in martial arts—something hinted at but never really directly addressed in the Iron Man comics. It is important that he can do this, though, as his suit often powers down or malfunctions and he loses his advanced weapons capabilities. He needs at least some grounding in this.

This was cleverly shown as part of the huge Marvel-wide "Civil War" epic. In the portion unfolding mostly in Iron Man (and collected in the 2007 graphic novel *Iron Man: Civil War*), Tony is shown in a flashback asking Captain America for some pointers on hand-to-hand fighting. This was meant to dovetail into the early years of the Avengers and clearly highlights the status Captain America has as a dominant hand-to-hand fighter in the Marvel Universe. The images in figure 9.2 are extracted from the "Iron Man: Civil War" story and show Tony learning (panel A) and receiving (panel B) various techniques from "Cap." Later, Tony shows that he has learned a bit from his training. I love how Cap congratulates Tony on getting it right and sweeping his legs out from under him. As Cap is falling to the floor, panel C shows him shouting, "Nice. Good follow through." When teaching martial arts this kind of thing actually happens all the time. I always let my trainees know if they are getting it right. It is kind of bizarre, though, because really you are congratulating them on throwing you down, intercepting your attack, or applying some kind of painful joint lock!

Anyway, using a framework and timeline I outlined in *Becoming Batman: The Possibility of a Superhero*, Tony would need at least the initial and middle training stages. I reckon he should train in some form of martial arts emphasizing striking, kicking, and joint locking (like Cap is doing to Tony's elbow in panel B) for five to eight years. He could do much of this training in parallel with his jet fighter qualifying. So, quite a bit of training is needed to be Iron Man. Or, check that. To be a successful Iron Man, a lot of training is needed!

All that extra training paid off handsomely for Tony in the story "The War with the Kree Is Over" found in the New Avengers: Illuminati comic miniseries that ran in 2007. In the miniseries, Iron Man, along with the rest of the Illuminati (such as Captain America, Professor Xavier, Reed Richards, Dr. Strange, Namor, and Black Bolt), is captured by the Skrull. In this particular story by Brian Michael Bendis and Brian Reed, Tony loses his armor to the Skrull and is awaiting his fate in a prison cell on board the ship. For some reason, the Skrull think it would be a neat idea to pretend to be other members of the Avengers (like Thor) to see what Tony would do. Because the Skrull are just Skrull and not super-powered Avengers, what Tony does is, well, kick butt and take names (although not so much on the names part). Once he has subdued all of the evil Skrull, he quips

A

Gross motor skills and simple moves are used.

B

Tony shown learning some basic fighting moves he could do with his armor on or off.

C

NICE. GOOD FOLLOW-THROUGH.

Eventually Tony shows he can display the benefits of his training. He needs at least five years of martial arts training for general body awareness and for use when the suit's other weapons are down or the suit has been taken away.

Figure 9.2. Tony Stark and Captain America training in martial arts from the graphic novel *Iron Man: Civil War* (2007). Note that Tony focuses on simple and effective techniques he could use while wearing his armor if his weapons systems failed. Copyright Marvel Comics.

"Thanks for the combat training, Cap." It is worth noting also that beyond the need for being able to fight in the Iron Man suit, the general body awareness that accrues with extended martial arts training would be very useful for keeping Tony Stark's nervous system highly "tuned" and ready to go. Robert Downey Jr., who has so capably played Tony Stark in the Iron Man film adaptations, has found martial arts training to be a useful and tuning influence in his own life. He has studied wing chun, a very direct and effective Chinese martial art, for many years.

What Is the Reality Check on Iron Man?

So, Iron Man as a possibility of a human machine certainly exists. However, the closest that we have today is more reflective of the original Iron Man introduced by Stan Lee back in 1963. The capabilities of that Iron Man are very similar to a combination of the Cyberdyne HAL and the fixed wing of Yves Rossy. Just as the Iron Man of almost 50 years ago is only now being realized, perhaps the fantastic seeming Iron Man of our day will be a reality 50 years from now.

In my view, exploring the bits of Iron Man that are realistic unveils the fantastic capacities the human body possesses. What is my verdict on the possibility of Iron Man? Well, it is possible if we accept a 1963 comic book version with some caveats. However, the length of Iron Man's career would be very short. (We will look at this more in detail later in the chapter.) Tony Stark as Iron Man faces three main problems in trying to have a long life as an armored superhero: concussion, the integrity of his neural implants, and the danger associated with experimenting on and implementing of the Iron Man suit. That is, Tony Stark would likely have killed himself in some accident or during flight. Also, given the need for direct nervous system connection, all the health issues associated with implanting man-made materials into the body would severely limit his ability to have a long career.

What Does Getting Whipped around by Whiplash Do to the Man within Iron Man?

Concussion is a huge issue for Tony to get past. Tony stated this very plainly in the 2010 graphic novel *Iron Man: The End*: "All those years trusting my armor to protect me, I forgot there was just a man inside. That pummeling took a toll. Degenerative nerve damage eating away my coordination like a boxer after too many shots to the head." Very graphic stuff, even for a graphic novel. And very true. Recently concussion in sports has received a lot of attention. Some years ago the NFL got quite serious about addressing concussion in football, particularly as it affected the marquee quarterback position. More recently the issue of concussion has—finally—received more attention in the high impact sport of ice hockey. There is still a long way to go when it comes to a widespread change in attitude.

I really enjoy hockey and football, and this past year I was watching Hockey Night in Canada on the Canadian Broadcasting Corporation (CBC's Hockey Night in Canada is the NHL equivalent of NFL's Monday Night Football). At the end of the Montreal Canadiens versus Toronto Maple Leafs game (if you're not familiar with ice hockey, you can think Washington Redskins and Dallas Cowboys to get the NFL equivalent, or Manchester United versus Liverpool for the English football equivalent, of the rivalry), one of the co-hosts in a between-period segment presented an overview of the night's major head injuries in a short video montage. Following this the host managed a kind of weak joke about squashed melons (or something similar to that—the key word was squash, it being Hallowe'en and visions of pumpkins were in the air I guess. Ha ha.). Another co-host laughed but clearly mostly out of pity, as he looked quite uncomfortable.

It made me wonder if we are so desperately out of touch with the grave consequences of neural damage that can be inflicted and evidenced by concussion that it remains the stuff of jokes. If so—and based on the evidence this is indeed yes—why is that? Is it simply because we cannot see it? What I mean is, clearly, we can see the effects of concussion. What we cannot see is the actual damage to the brain, the changes in energy demand and energy delivery and the disordered activity of the neurons. A schematic illustration of what happens with a concussive event is shown in figure 9.3. Panel A of the figure shows the head impacting an object and the motion of the brain within the skull. When the head hits a solid object, the brain moves forward and then contacts the inside of the skull. This is the case whether the head (or helmet of Iron Man) hits a solid object or is hit by something. The relative motion is the same and gives rise to the mechanical impact of the brain. A head-first incident like the one shown in the figure is experienced first on the "front" of the brain (frontal cortex) and then secondarily on the back of the brain (occipital cortex). The second impact is called "contre coup."

Strong head trauma provokes a cascade of events that leads to an energy crisis—kind of a malevolent neuronal oxygen debt—that causes the neurons in the brain to fail. This is shown in panel B of figure 9.3. The increase in energy demand, coupled with reduced blood flow and reduced metabolism, leads to the death of some nerve cells. Neurons will also take several days to recover from this massive shift of activity. The memory problems and mental fuzziness and

A

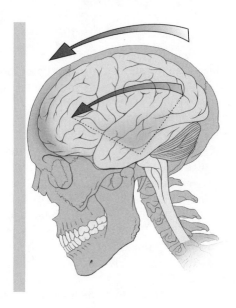

B

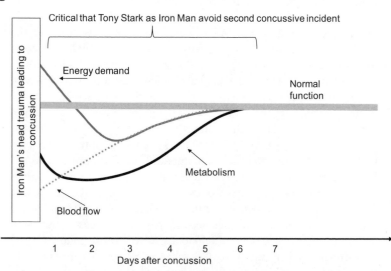

Figure 9.3. A schematic of what happens with a concussive event. The head impacting an object and the motion of the brain within the skull (A). When the head hits a solid object, the brain moves forward and then contacts the inside of the skull. The effects of concussion on energy demand, metabolism, and blood flow in the brain (B). The key period when Tony must avoid additional concussive event is indicated at the top of the panel. Panel A image courtesy of Patrick J. Lynch; panel B data redrawn from Shulman (2000).

confusion seen during this time after concussion occur because of what has happened at the cellular level. Returning to normal concentrations of neurotransmitters can sometimes take up to two weeks.

I can personally vouch for these effects due to an accident I had during the editing of this book. I suffered a "mild" concussion and for about ten days it felt like I was two steps behind everything that was happening to me. I was exhausted, had difficulty concentrating, and generally was out of sorts. Then, it slowly went away. It felt like a mist being burned off by sunshine. I felt better and better. But it took those many days for all those ions and neurotransmitters to get back to their appropriate levels.

Some of the main areas of the brain that are most affected by repetitive mild head trauma are the hippocampus as well as the frontal lobes. These areas are very important for memory formation and storage and for movement control. Unfortunately, these brain areas are also particularly sensitive to the large changes in gravitational forces that occur during the rotation of the head in a concussive event. Rest assured that Iron Man in action experiences "concussive events" on a daily basis.

This type of injury is a real occupational hazard in contact sports like hockey and football. In the NHL another high profile incident happened in 2011. The league's star player Sidney Crosby of the Pittsburgh Penguins received a strong shoulder (and hard plastic shoulder pad) in a blind-side hit. He was knocked down and looked a bit unsteady and concussed as he stood up and skated back to the bench. He was cleared to return to play. Then, six days later, he was hit against the boards and suffered another concussive incident. He was knocked out of playing for the entire season.

Concussion really is a hot button topic. In 2009, the Associated Press published the results of an informal survey of NFL players conducted in November of that year. The AP surveyed 160 NFL players about their experiences with concussion. The survey included a mix of rookies to 17 year veterans and those playing all positions. Thirty of those players, that is, just under 20% of those interviewed, revealed that they had either not disclosed or trivialized their own concussions. Additionally, half of the players indicated they had experienced a concussion, and just over one-third of the players said the concussion had forced them to miss playing time. This information matches that from the Canadian Football League in 2000, where approximately 50% of players indicated they had experienced a concussion.

It is now much more widely understood that the brain has tremendous adaptive abilities. The nervous system really does have a "plastic" ability to respond to training or to compensate for damage. This is critical because, when we are talking about concussion, the main point that must be understood is "compensation," which is really another way to say "repair." A good repair means things still work well but does not mean things are the same as before the repair was either needed or finished. There is only so much repairing that can go on until limitations arise. These limitations should not be ignored or celebrated in Iron Man's world or in our own. Also, since Iron Man is repeatedly battling multiple foes and being blasted more than once, we need concern over repeated concussion.

When someone hasn't yet fully recovered from the initial concussion and then suffers another blow to the head, a very dangerous "secondary impact" syndrome can develop. Then, even a minor impact can trigger a further and more dramatic change in the regulation of blood supply to the brain. This response can lead to swelling and blood pooling and very often is fatal. A less severe outcome of multiple concussions can occur even when there has been recovery after each concussion but the concussive impacts are repeated. Often this is called "post-concussion syndrome" but anecdotally most of us know it as being "punch drunk." It was historically named this because it was noticed initially in boxers who do absorb many blows to the head. As Tony Stark remarked above in the quote from *Iron Man: The End*, he really has had significant damage to his nervous system from all the trauma he has experienced. Your brain has tremendous capacity for change and recovery, but eventually those limits will be exposed.

Clearly, concussion and brain trauma is a constant danger for Iron Man. This is not so much from being body checked or tackled—although that would still contribute. Especially being tackled or checked or bashed by Whiplash or Iron Monger—but more being bashed into things. Or even surviving the g-forces associated with taking off and flying around. Rather, in the world of Iron Man he is constantly being bombarded and blasted. In fact, this is in a story found in the 2007 *Iron Man: Hypervelocity* graphic novel. Tony talks about how harsh the effects on the body are when using his older suits compared to his newer versions: "No more soft, wet, organic brain sloshing merrily against the inside of a bony cranial vault means no more troublesome concussions. Also, a welcome respite from post-concussion vomiting into my helmet."

To be honest, the structure of the Iron Man suit doesn't provide much protection against concussion for the head. Have a peak back at the thin face plate shown in the action figure in figure 1.4. This wouldn't be very helpful in reducing impact forces. As shown at the bottom of figure 9.3, it is important Tony Stark not receive another concussive incident for at least a week to avoid potential serious injury. However, this would be hard to avoid. For example, during the climactic battle with Iron Monger near the end of the 2008 *Iron Man* movie, I estimate that Iron Man receives seven concussive events within the span of one minute. Forget about resting between concussions for seven days! Those concussive events included slamming into a car, being pummeled by Iron Monger, being slammed into a bus, and a concussive event from a blast. Let's return to the bomb suit concept and blast injury.

Here I want to focus specifically on the primary blast wave and how it affects the brain, particularly on the impact of blasts on military personnel. (You can refer back to figure 7.2 for an illustration of the extreme pressurization and impact.) Ibolja Cernak and Linda Noble-Haeusslein have conducted some excellent work in this area, which is also known as "blast-induced neurotrauma." This is a very important field of study, given the steadily increasing numbers of military personnel suffering this kind of injury. It has been estimated that almost three-quarters of the U.S. military casualties from Operation Enduring Freedom in Afghanistan and Operation Iraqi Freedom were caused by explosive weapons. Additionally, civilians are also routinely injured in blast-induced accidents of a similar kind. For the military personnel, protection from the blasts themselves would be a key advance. We touched on this briefly when discussing bomb disposal suits. In that context, protective suits are very useful for guarding against shrapnel. However, here we want to talk about the long-term effects of using an Iron Man suit in combat and being subject to multiple blast incidents. The protection from shrapnel using body armor may actually make the effect of the blast wave worse! The high density and rigid body armor can act as a good interface for transferring the energy of the blast wave to the wearer and by concentrating the blast energy as it moves into the body.

While this may seem confusing, think of an example using sound waves in music. Imagine listening at a door outside a room where loud music is being played. If the door is a hard surface (like a normal door), you could press your fingertips to the door and clearly feel the vibrations. The rigid door transfers the kinetic energy of the sound waves into vibration you feel with your fingers. If the door is covered with soft rubber and you try the same thing, you won't feel much at all. The soft rubber isn't a good conductor in this case. So, it shouldn't be that surprising that while penetrating injuries are reduced with body armor, blast wave injuries have increased.

Damage to the nervous system from the blast wave occurs in several different ways. When the blast wave arrives, it may move directly through the skull and cause a rapid rotation of the head. This is similar to the general mechanism of injury in a concussion. Moving at about the speed of sound, kinetic energy in the blast wave can be transferred to the fluid component of the body in the main blood vessels in the trunk, where it is then transferred to the nervous system. This action results in damage and destruction of neurons as well as interruption in the functions of the neurons and connections in the brain and spinal cord.

Based on work in many different species from rat to monkey, blast injuries that are non-fatal decrease ability to perform work or exercise, reduce hunger and appetite, can induce spasm of the blood vessels, reduce brain activity including memory and the ability to perform movement, and can cause swelling in the brain. These effects continue to be amplified by repeated exposure to new blast events. In many ways you can think of this as being like the cumulative effects of concussion that we talked about above.

It is clear from this that for Iron Man to have a long and neurologically intact career he must limit his blast exposure to almost nothing. A simple reading of any Iron Man comics or viewing of *Iron Man* or *Iron Man 2*, show this clearly is not the case. A good analogy is one I often use when describing physiological systems. If we compare your body to a computer operating system, with rare exceptions your physiology is a Macintosh, not a PC. With a PC you tend to find out and experience many little crashes and some big crashes too. With a Mac you are typically spared the little crashes and small errors are compensated for and hidden from you the user. That is, until they become too big. At that time, the Mac crash is humungous and can be catastrophic. Your body is like that in the sense that your systems adapt to stresses and compensate for damage until you are stretched very thin. Then catastrophic failure in the body can occur.

Integrity of Implants in Iron Man

Please think back to the revision of the origin story I suggested for Tony (have a peek back at chapter 7). Along with this revision is the corresponding implication that the use of the implantable cardioverting defibrillator (ICD) has for his activity as Iron Man. One of the controversies around ICDs, though, is how active people should be if they have an implantation. Typically, aggressive contact sports or vigorous exercise are not recommended for those with ICDs. There is concern over how well they would function in such cases and concern over inappropriate shocks occurring. That means that the function of the defibrillator itself could interrupt normal heart rhythm during extreme activity and lead to more significant problems. So, the upshot of all of this is that Tony Stark would likely have a difficult and dangerous time being Iron Man. He is clearly subjected to very vigorous exercise and extreme contact—for example getting smashed around by Iron Monger as in the first *Iron Man* movie. This would definitely lead to the possibility of inappropriate shocks, or, more importantly, increased breakdown of the electrical leads used to provide the shocks for restoring heart rate.

Normally, there can be a steady increase in failure rate and problems eight to ten years after implantation of a cardioverting defibrillator. This means the best case is that Tony may well need another implant after about a decade. However, his work as Iron Man clearly isn't the "best case" for long-term stability and integrity of the ICD. How much it would be shortened is almost impossible to determine, because the data aren't really available on how the kind of violent activity that Iron Man experiences may affect the lifespan of ICDs. However, there is some research on the need to replace cochlear implants for hearing restoration. After an implantation of a cochlear implant, current technology only sees a 5% chance of needing a replacement (called "revision" or "reimplantation"), which seems pretty good. Of interest and very relevant for what we have just been talking about, is that almost half of the cases where hardware failure has led to a revision surgery have involved a history of head trauma! So, all the bashing around we were just talking about clearly is quite hard on brain-machine interface.

Another consideration is the integrity of the nervous system and the electrode connection that would be necessary for the full integration of the suit needed for realistic use. It should be remembered that it is very difficult to interface technology with biology. Based on state-of-the-art experiments with different animals, it seems only about 50% of implanted electrodes can be usefully employed. Of those that could be used, there is a significant reduction in "usefulness" over time. This is largely due to a process known as encapsulation or "reactive gliosis"—a type of scarring of the nervous system—associated with immune rejection. Think of the medical intervention needed to keep the body from rejecting a transplanted organ. Now imagine that instead of an organ a piece of machinery has been implanted into the brain.

Jennie Leach and colleagues describe the view that the bodily response to an implanted neuroprosthetic interface has a rapid acute response, characterized by injury, inflammation, and what is known as "microglial activation." Nerve cells are neurons and other stuff. In the other stuff category we find glial cells. Microglia make up about 20% of glial cells in the brain and provide the main protective immune response cells in the brain and spinal cord. They are a kind of "macrophage," which means they attack and digest invaders and foreign objects in the nervous system and are the first and main form of active immune response in the brain and spinal cord.

The immune response is typically to digest an intruder and, if it cannot be digested, to cover it up so that it can do no harm. Implanted electrodes cannot be easily destroyed so the cover-up process is instead the main outcome. This begins in the acute phase and continues in a chronic response, which results in the formation of a virtually impenetrable glial or fibrotic scar around the implant. An example of this cascade of events is shown in figure 9.4. Panel A shows the implantation of an electrode array ("Utah array") inserted through the brain and implanted on the surface of the brain. The drawings in panels B through D are close-ups of the region right beside the implant (shown as dashed rectangle in A). Cellular organization in the cortex is shown prior to implantation (panel B), immediately after implant ("acute," panel C), and in chronic implantation (panel D). The key point is that there is steadily increasing scarring of support neurons (glia), shown at the far left near the implant surface, and some death of neurons, largely related to inflammatory responses.

This cascade is also of concern for other kinds of neuroprostheses like retinal implants for vision and cochlear implants for hearing. An interesting approach to deal with the significant scarring problem for cochlear implants has been to use pharmacological treatments to

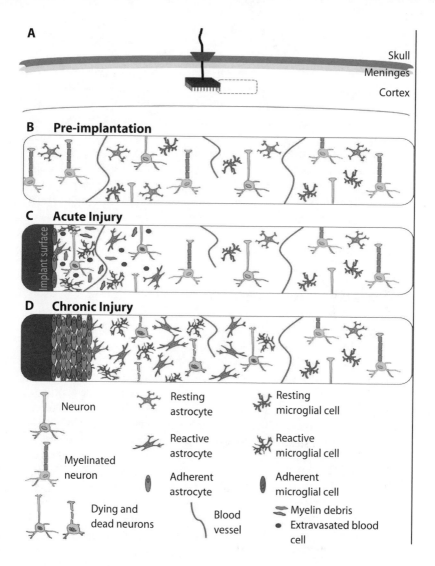

Figure 9.4. An example of the cascade of events occurring in the brain after implantation of an electrode array for brain-machine interface. The implantation of an electrode array, known as a Utah array, on the surface of the brain (A). Close-ups of the region right beside the implant (B–D; shown as dashed rectangle in A). Cellular organization in the cortex is shown prior to implantation (B), immediately after implant ("acute," C), and in chronic implantation (D). Note steadily increasing scarring of support neurons (glia), shown at the far left near the implant surface, and some death of neurons, largely related to inflammatory responses. Courtesy Leach (2010).

help trick the nervous system. One such treatment that has shown some promise is to use brain-derived neurotrophic factor (BDNF), which is a major player in a family of chemicals that help in neural development and neural plasticity. In order to avoid the problem of scarring, tissue engineers and nanotechnology experts are working on ways to make the implant appear more biological to the body using means such as this. Even though this pharmacological approach has promise, there is a long way to go to get to a stable enduring implant like that needed for a full brain-machine interface in an Iron Man suit. And then, on top of that, to have the brain subjected to the kind of continual trauma that Tony would experience isn't really a recipe for long-term success.

Training Keeps Tony's Brain from Getting Rusty

A big part of thinking about Iron Man's career is his ability to work against the normal decline of nervous system function that happens with aging. Maybe there is some protection from changes in the aging nervous system due to all his training? Biological aging—senescence—is a steady and inevitable process. Senescence captures the reduction in function that starts just around the third decade in humans. Good examples of aging-related changes in the nervous system are those in the motor system. As part of the motor unit concept we talked about earlier, recall that motor neurons in the spinal cord send out their axons to connect with the fibers in your muscles. Well, with aging motor neurons die. Since the motor neurons are the final relays for sending the commands to make muscle contract, it is reasonable to think that muscle strength would decline.

Between the ages of 20 to 90 years, half of the skeletal muscle mass can be lost and this causes a corresponding reduction in muscle strength. Muscle fibers also will die and in particular those in the fastest contracting motor units, the type IIa group. Despite all that going on, you wouldn't notice the reduction in motor neuron number and muscle fibers that much, because your nervous system has a great ability to cover up problems. Plasticity in the nervous system compensates for the death of muscle fibers and motor neurons in senescence in a very clever way. A large muscle in your leg could have 250 motor units in it, and each unit might connect to a thousand muscle fibers. The distribution is called the "innervation ratio," and, in this example, would be 1,000. By the time Tony Stark hits age 70, the number of his motor units might drop from 250 to 125 in this muscle. So, he would have lost 125 motor neurons. However, many of his muscle fibers (previously connected to those now-dead motor neurons) would remain alive. They would sit in his muscle waiting to do their normal jobs, and this would happen when other motor neurons innervating muscle fibers in that leg muscle send branches from their axons over to the muscle fibers that are now "disconnected" from their original neurons. This process is called "sprouting" and is similar to the process of sensory reinnervation we talked about earlier with the face transplantation. Overall, this process of sprouting creates much larger innervation ratios—in this example let's say it is now 1,500 fibers per neuron—and helps maintain strength as you age. This is very similar to what occurs in recovery from some nerve injuries and is affected by how much activity the nervous system sees. So, if Tony is very physically active, his ability to maintain the integrity of his motor system in this way will be improved.

We can extend the example of Tony's motor system to many other parts of his nervous system. The nervous system is just like all other systems and parts of your body. When not used repeatedly, your physiological systems try to be as efficient as they can be. The nervous system responds to stresses to minimize the effect. You have experienced this throughout your life probably without really paying much attention to it. If you have ever done some strength training, you have a good specific example in mind. Imagine doing a bunch of arm flexion exercises ("biceps curls"). The stress on the muscles doing those curls leads to an increase in strength. Now imagine you did a bunch of training and got stronger but then couldn't train for a while because you broke your arm (sorry—it is just an example. Nothing personal). Suddenly you wouldn't be able to use your one arm very much and it would get weaker. You removed the stress now, so your body doesn't maintain the muscle to the same level. However, because of your training, you had built up a bit of a reserve of muscle strength. You still did weaken when you weren't able to use your arm, but, the really important part is that you would still finish stronger than if you hadn't done the training in the first place. So, the training created a kind of reserve that helped buffer the lack of use that occurred later.

Well, you don't actually curl weights with your brain. But the command to do the curling comes from your brain. Tony Stark is stimulating his brain by straining and training every time he does physical activity or interfaces with his Iron Man suit. He places a lot of extreme demands on his nervous system and creates a kind of "brain reserve" by doing this. Exercise can also stimulate the brain to not only maintain the neurons and synaptic connections but also to create new neurons. Since declining function in the nervous system is inevitable as Tony gets older (and which was really well described in the graphic novel *The End* from 2010), any reserve he can add will keep a higher function as he gets older.

Brain reserve describes the idea that larger brains with more neurons and more synaptic contacts might be better at dealing with problems during aging, for example, dementia and disorders like Alzheimer's disease. Cognitive reserve is specific to how brain reserve can affect the ability to think and reason when there is pathological damage to the brain. So, the more reserve you have means the less you will be affected by declines in function. This doesn't mean Tony Stark could stop the normal inherent decline in function. However, it does mean that he can reduce the impact of the decline. That old phrase of "use it or lose it" applies really well to all aspects of your nervous system!

How Long Will the Iron Avenger Last?

Let's assume that Tony avoids concussion (or at least too many of them), the bodily rejection of his brain-machine interface, and serious injury from falls and weapons blasts. Let's also assume that his heart remains healthy and that he keeps his mind in shape to be able to multitask efficiently. What then? When should he fly off to the retirement home for heroes?

In his origin story, we learn that Tony took control of Stark Industries at the age of 21 and that shortly after that had his little "incident" with some shrapnel that put him on the road to inventing Iron Man. Remember that Tony will take more than ten years to go through pilot and hand-to-hand combat training. Remember too that it took Rossy, Nuytten, and Sankai decades to modify their suits and that we allowed 40 years for Tony to go through that process.

If my math is correct, Tony will be in his 60s by the time he masters being Iron Man. Those of you who read *Becoming Batman* will recall that, using sports icons as a guide, we determined that Batman would have to hang up his cape in his mid-50s, at the absolute latest. So at some point Tony may want to start inventing suits for younger crime fighters to wear. A big caveat, though, is that the estimates for all these timelines are based on thinking of pretty linear and steady progress towards the ultimate objective of Iron Man. But advances in science and engineering don't always work like that. There can be wholesale paradigm shifts (like those Thomas Kuhn talked about earlier) and suddenly fields can move forward in big jumps. For Tony Stark's sake (and for our own imaginations), let's assume some of those come along during his years of Iron Man research and development. I feel better thinking of Tony as both inventor and user— however short that career as user might be.

The End

The last point of this chapter is that for a real Iron Man suit of armor to exist and to be usefully applied, it must be based on the kind of brain-machine interface hinted at in the Extremis story line. It must respond without having to be consciously commanded. This would free up the neurological resources needed for more challenging tasks and environments. The user must be highly trained and in tip-top shape. Remember, the suit amplifies the user. If the user is "poor quality," you just get "louder" poor quality when it is amplified. This was pointed out by Whiplash (Ivan Vanko) in *Iron Man 2*. He was brought in by Justin Hammer to create an army (and navy, and air force, and marines!) of "Iron Man–like" suits for soldiers to wear. But instead he creates remote-controlled robot drones. He tells Hammer "Drone better . . . human causes problems." Getting around those problems requires a lot of work still.

We are on a line of discovery that may well one day produce what was written on the Invincible Iron Man masthead beginning with issue #70 in September 1974: "When millionaire industrialist Tony Stark, inventor extraordinaire, garbs himself in solar-charged, steel-mesh armor he becomes the world's greatest human fighting machine." Tony's father, Howard Stark, said it best in one of the old Super 8 film montages shown in *Iron Man 2*: "everything is achievable through technology." The fields of neuroscience and biomedical engineering continue to lead efforts to arrive at a useful working concept like that of an Iron Man neuroprosthetic. We aren't there yet. And certainly the jet boots and repulsor rays are not even on the radar. But much of what we have discussed is on the horizon. We look to the efforts of scientists, engineers, and inventors to continue to take us along that path. The real-life Invention of Iron Man lies ahead.

Appendix

TEN MOMENTOUS MOMENTS OF
THE METAL MAN

Iron Man comics have been divided (so far) into five volumes. This does not include the original Iron Man stories in Tales of Suspense. As detailed in Marvel Comics' *Iron Man: The Official Index to the Marvel Universe* (2010), Iron Man showed up in Tales of Suspense from his debut in issue #39 from 1963 up until issue #99 in March 1968. Then Iron Man debuted in his own comic from Iron Man #1 in May 1968. This "volume 1" of Iron Man was maintained until issue #332 in September of 1996. Volume 2 spanned Iron Man #1 in November 1996 until #13 in November 1997. Volume 3 began with Iron Man #1 (yes, this is a little confusing) in February 1998 until #89 in December 2004. Volume 4 began as Iron Man #1 (again!) in January 2005 and ran until #32 in October 2008 (although it was called Iron Man: Director of S.H.I.E.L.D. from issue #15). Volume 5 started as The Invincible Iron Man #1 in July 2008. In January 2011, the numbering became consecutive, beginning with #500 to mark the five hundredth issue of the comic. There have also been additional offshoot titles (such as War Machine, which also has several volumes) and special collections across the years.

The ten momentous moments of Iron Man are listed in the table following.

Iron Man era	Date and reference	Event description and relevance
Tales of Suspense	#39 from March 1963 "Iron Man Is Born!"	Debut of Iron Man in a story penned by Stan Lee and Larry Lieber with art by Don Heck and lettering by Art Simek. This is the first Iron Man origin story with Tony Stark injured in Vietnam and forced (along with Ho Yinsen) by Wong-Chu to build weapons. Instead they build the original Iron Man "gray armor." My favorite quote is Tony Stark to Wong-Chu: "You are not facing a wounded dying man now—or an aged, gentle professor! This is Iron Man who opposes you and all you stand for!"
	#48 from December 1963 "The New Iron Man Battles . . . The Mysterious Mr. Doll!"	Tony creates a new red and gold armor system that is much reduced in bulkiness. It is a much more anthropomorphic suit and is shown as easily modular and storable. While it looks much more awe-inspiring, it is less safe and protective against concussive impacts than the old armor.
Iron Man (Volume 1)	#200 from November 1985 "Resolutions!"	This marks the first time that the Iron Monger armor created by Obidiah Stane's company is shown in the comics. The Iron Monger armor is much larger and bulkier than even the original gray armor and was built from Stark's old notebook plans. The Iron Monger armor is highly protective and is closest to what is currently available for "driveable" exoskeletal armor.
	#284 from September 1992 "Legacy of Iron"	This is the full debut of Jim Rhodes as War Machine. Tony (in a "posthumous" transmission) tells Rhodey he created the new heavily militarized armor specifically for him and his body. This hints at the need for customized armor and the fact that a neuroprosthetic as advanced as the Iron Man and War Machine suits cannot simply be worn like clothing off a rack.
	#290 from March 1993 "This Year's Model"	Tony is partially paralyzed and it is uncertain whether he will be able to regain movement. So he creates a remote control "telepresence armor" (NTU-150) that makes use of a "telepresence" headset. The general idea presages concepts of brain-machine interfaces being explored now.

Source	Description
Iron Man (Volume 4) — #1–6 from January 2005 to May 2006 "Extremis" story arc	First description of the "Extremis" neural interface armor written by Warren Ellis with art by Adi Granov. Iron Man's origin story is updated from Vietnam to the Gulf War and the concept of a neural interface for the Iron Man suit is described. This concept closely approaches the way in which the Iron Man suit would have to function, that is, as a fully integrated neuroprosthetic.
#1–6 from March to August 2007 in special "Iron Man: Hypervelocity" series	In "Hypervelocity," the Iron Man armor gains sentience and goes berserk. It flies around without a "pilot" wreaking havoc until Tony, using older armor, eventually defeats it. A good warning of the balance between needing to create almost autonomous armor that can be controlled by a human user with safeguards (think Isaac Asimov's 3 laws of robotics) against going out of control.
Iron Man (Volume 5) — Invincible Iron Man #1 from July 2008 "The Five Nightmares, Part 1: Armageddon Days"	This marks the debut of Matt Fraction's tenure as writer of Iron Man. This storyline uses concepts from the neural interface of "Extremis" with "biological upgrading" of humans and direct nanotech interface.
Marvel Studios Films — Iron Man (2008)	This movie introduced the motorized robotic equipment for dressing Tony Stark in Iron Man armor. Right now we don't even have the technology to safely dress Tony Stark in the Iron Man as shown in the robotics, let alone create the Iron Man armor itself.
Iron Man 2 (2010)	At his birthday party, Tony Stark is shown wearing and using the Iron Man suit while heavily intoxicated. Jim Rhodes dons the spare (soon to become War Machine) armor to try and subdue him. Shows very clearly the horrific implications of mixing technology and alcohol abuse.

Bibliography

For further reading about the realities of superheroes, have a look at *Becoming Batman: The Possibility of a Superhero* (Johns Hopkins University Press, 2008). *Physics of Superheroes* by friendly-neighborhood physics professor James Kakalios (Gotham, 2009) is a great book. If you want further information on Iron Man, I suggest *Iron Man: Beneath the Armor* by Andy Mangels (Del Rey 2008), *Iron Man: The Ultimate Guide to the Armored Superhero* by Matthew Manning (Doring Kindersley, 2010), or *Marvel Chronicle* by Tom DeFalco, Peter Sanderson, Tom Brevoort, and Matthew Manning (Doring Kindersley, 2008). An excellent exploration of Iron Man in pop culture is *Comic Book Nation* by Bradford Wright (Johns Hopkins University Press, 2003). For more on neuroscience, I suggest checking out "Brain Facts," a primer on the nervous system published and made freely available at www.sfn.org by the Society for Neuroscience. More on brain-machine interface can be found in *Beyond Boundaries* by Miguel Nicolelis (Times Books, 2011).

Comics and Graphic Novels Cited

1963   Tales of Suspense #39: "Iron Man Is Born!"
        Tales of Suspense #40: "Iron Man vs. Gargantus"
        Tales of Suspense #48: "The New Iron Man Battles . . . the
            Mysterious Mr. Doll!"

1972   Invincible Iron Man #47: "Why Must There Be an Iron Man?"

1974   Invincible Iron Man #70: "Who Shall Stop Ultimo?"

1979   Invincible Iron Man #121: "A Ruse by Any Other Name"

Invincible Iron Man #122: "Journey!"
Invincible Iron Man #123: "Casino Fatale!"
Invincible Iron Man #124: "Pieces of Hate!"
Invincible Iron Man #125: "The Monaco Prelude"
Invincible Iron Man #126: "The Hammer Strikes"
Invincible Iron Man #127: ". . . A Man's Home Is His Battlefield!"
Invincible Iron Man #128: "Demon in a Bottle"
Invincible Iron Man #129: "Dreadnight of the Dreadnought!"

1980   Iron Man #142: "Sky Die"
        Iron Man #144: "Apocalypse Then"

1981   Iron Man #150: "Knightmare"

1983   Invincible Iron Man #169: "Blackout!"
        Invincible Iron Man #170: "And Who Shall Clothe Himself in Iron?"

1984   Invincible Iron Man #182: "Deliverance"

1987   Iron Man #218: "Deep Trouble"

1989   Iron Man #242: "Master Blaster"
        Iron Man #243: "Heartbeaten"
        Iron Man #244: "Yesterday . . . and Tomorrow"
        Iron Man #245: "Inside Angry"
        Iron Man #249: "The Doctor's Passion"

1992   Invincible Iron Man #280: "Technical Difficulties"
        Invincible Iron Man #281: "The Masters of Silence"
        Invincible Iron Man #282: "War Machine"
        Invincible Iron Man #284: "Legacy of Iron"
        Invincible Iron Man #286: "Dust to Dust"

1993   Invincible Iron Man #290: "This Year's Model"
        Invincible Iron Man #291: "Judgement Day"

2005   *Ultimate Iron Man* (graphic novel)
        Invincible Iron Man: Extremis, Parts 1–4

2006   Invincible Iron Man: Extremis, Parts 5–6

2007   The Invincible Iron Man #10: "Execute Program, Part 4"
        Iron Man: Civil War (graphic novel)
        Iron Man: Extremis (graphic novel)

Iron Man: Hypervelocity, Parts 1–6
Iron Man / Captain America: Casualties of War #1:
  "Civil War—Rubicon"
Iron Man: Director of S.H.I.E.L.D. #31: "With Iron Hands, Part 3"
Iron Man: Director of S.H.I.E.L.D. #33: "War Machine, Part 1:
  Weapon of Shield"
New Avengers: Illuminati #1: "The War with the Kree Is Over"

2008   The Invincible Iron Man #1:"The Five Nightmares,
           Part 1: Armageddon Days"
        The Invincible Iron Man #2: "The Five Nightmares,
           Part 2: Murder Inc."
        The Invincible Iron Man #3: "The Five Nightmares, Part 3: Pepper
           Potts at the End of the World"
        Invincible Iron Man: The Many Armors of Iron Man (graphic novel)
        Iron Man: Demon in a Bottle (graphic novel)
        Iron Man: Director of S.H.I.E.L.D.—Haunted (graphic novel)
        Iron Man: Iron Manual (graphic novel)
        Secret Invasion (graphic novel)
        War Machine (graphic novel)

2009   Invincible Iron Man #8: "World's Most Wanted, Part 1:
           Shipbreaking"
        Invincible Iron Man #9: "World's Most Wanted, Part 2: Godspeed"
        Invincible Iron Man #13: "World's Most Wanted, Part 6: Some King
           of the World"
        Invincible Iron Man #14: "World's Most Wanted, Part 7: The Shape
           of the World These Days" sometimes parts have separate names
           and sometimes not
        Iron Man Armor Wars #1: "Down and Out in Beverly Hills"
        Iron Man Armor Wars #2: "The Big Red Machine"
        Iron Man Armor Wars #3: "How I Learned to Love the Bomb"
        Iron Man: Requiem (graphic novel)

2010   Iron Man Armor Wars #4: "The Golden Avenger Strikes Back"
        Iron Man: The End (graphic novel)

MOTION PICTURES AND TELEVISION PROGRAMS CITED

Avatar (2009, 20th Century Fox)
Doctor Who (1963–1989, 2005–present; BBC TV series)
Iron Man (2008, Marvel Studios)
Iron Man 2 (2010, Marvel Studios)

The Six Million Dollar Man (1974–78; ABC TV series)
Surrogates (2009, Touchstone)
The Terminator (1984, MGM)

BOOKS AND JOURNAL ARTICLES

Preface

Mangels, A. (2008) *Iron Man: Beneath the armor*. Del Ray Books, New York.

Chapter 1. Origins of the Iron Knight: Bionics, Robotic Armor, and Anthropomorphic Suits

de La Hire, J. (2009) *Enter the Nyctalope*, translated by Brian Stableford. Hollywood Comics, Encino, CA.
Mangels, A. (2008) *Iron Man: Beneath the armor*. Del Ray Books, New York.
Marvel. (2008) *The official handbook of the Marvel Universe*, vol. 2., master edition. Marvel Publishing Inc., New York.

Chapter 2. Building the Body with Biology: When the Man of Metal Needs to Muscle In

Biddiss, EA, and Chau, TT. (2007) Upper limb prosthesis use and abandonment: A survey of the last 25 years. *Prosthetics and Orthotics International* 31(3): 236–57.
Burkett, B. (2010) Technology in Paralympic sport: Performance enhancement or essential for performance? *British Journal of Sports Medicine* 44(3): 215–20.
Camporesi, S. (2008) Oscar Pistorius, enhancement and post-humans. *Journal of Medical Ethics* 34(9): 639.
Childress, DS. (2002) Development of rehabilitation engineering over the years: As I see it. *Journal of Rehabilitation Research and Development* 39(6 Suppl): 1–10.
Clower, WT. (1998) The transition from animal spirits to animal electricity: A neuroscience paradigm shift. *Journal of the History of Neuroscience.* 7(3): 201–18.
Ferris, DP, Bohra, ZA, et al. (2006) Neuromechanical adaptation to hopping with an elastic ankle-foot orthosis. *Journal of Applied Physiology* 100(1): 163–70.
Ferris, DP, Louie, M., et al. (1998) Running in the real world: Adjusting leg stiffness for different surfaces. *Proceedings of the Royal Society of London (Section B, Biologicial Science).* 265: 989–94.
Grabowski, AM, and Herr, HM. (2009) Leg exoskeleton reduces the metabolic cost of human hopping. *Journal of Applied Physiology* 107(3): 670–78.

Herschbach, L. (1997) Prosthetic reconstructions: Making the industry, re-making the body, modelling the nation. *History Workshop Journal* (44): 23–57.

Hobson, DA. (2002) Reflections on rehabilitation engineering history: Are there lessons to be learned? *Journal of Rehabilitation Research and Development* 39(6 Suppl): 17–22.

Marvel. (2008) *The official handbook of the Marvel Universe*, vol. 2, master edition. Marvel Publishing Inc., New York.

Michaelis, V. (2010) Runner lives life without limitations. *USA Today*, April 26, 2010, 2.

Moritz, CT. (2009) A spring in your step: Some is good, more is not always better. *Journal of Applied Physiology* 107(3): 643–44.

Muri, A. (2007) *The enlightenment cyborg: A history of communication and control in the human machine*. University of Toronto Press, Toronto, Canada.

Noth, J. (1992) Motor units. In *Strength and power in sport*, edited by PV Komi, Blackwell Scientific Publications, London.

Parsons, P. (2010) *The science of Doctor Who*. Johns Hopkins University Press, Baltimore.

Thurston, AJ. (2007) Paré and prosthetics: The early history of artificial limbs. *Australia and New Zealand Journal of Surgery* 77(12): 1114–19.

Vogel, S. (2001) *Prime mover: A natural history of muscle*. Norton, New York.

Warwick, K. (2002) *I, cyborg*. University of Illinois Press, Champaign.

Weyand, PG, Bundle, MW, et al. (2009a) The fastest runner on artificial legs: Different limbs, similar function? *Journal of Applied Physiology* 107(3): 903–11.

Weyand, PG, Bundle, MW, et al. (2009b) Point: Counterpoint: Artificial limbs do/do not make artificial running speeds possible. *Journal of Applied Physiology* 108: 1111–12.

Chapter 3. Accessing the Brain of the Armored Avenger: Can We Connect the Cranium to a Computer?

Allison, BZ, Wolpaw, EW, et al. (2007) Brain-computer interface systems: Progress and prospects. *Expert Review of Medical Devices* 4(4): 463–74.

Andersen, RA, and Cui, H. (2009) Intention, action planning, and decision making in parietal-frontal circuits. *Neuron* 63(5): 568–83.

Andersen, RA, Hwang, EJ, et al. (2010) Cognitive neural prosthetics. *Annual Review of Psychology* 61(1): 169–90.

Buschman, TJ, and Miller, EK. (2007) Top-down versus bottom-up control of attention in the prefrontal and posterior parietal cortices. *Science* 315(5820): 1860–62.

Fernandez, E, Pelayo, F, et al. (2005) Development of a cortical visual neuroprosthesis for the blind: The relevance of neuroplasticity. *Journal of Neural Engineering* 2(4): R1–R12.

Hochberg, LR, Serruya, MD, Friehs, GM, et al. (2010) Neuronal ensemble control of prosthetic devices by a human with tetraplegia. *Nature* 442: 164–71.

Hwang, EJ, and Andersen, RA. (2009) Brain control of movement execution onset using local field potentials in posterior parietal cortex. *Journal of Neuroscience* 29(45): 14363–70.

Leach, JB, Achyuta, AKH, et al. (2009) Bridging the divide between neuroprosthetic design, tissue engineering and neurobiology. *Frontiers in Neuroengineering* 2.

Millett, D. (2001) Hans Berger: From psychic energy to the EEG. *Perspectives in Biology and Medicine* 44(4): 522–42.

Nicolelis, M. (2011) *Beyond boundaries: The new neuroscience of connecting brains with machines—and how it will change our lives.* Times Books, New York.

Normann, RA, Greger, B, et al. (2009) Toward the development of a cortically based visual neuroprosthesis. *Journal of Neural Engineering* 6(3): 035001.

Patil, PG, and Turner, DA. (2008) The development of brain-machine interface neuroprosthetic devices. *Neurotherapeutics* 5(1): 137–46.

Penfield, W, and Rasmussen, T. (1950) *The cerebral cortex of man: A clinical study of localization of function.* Macmillan, New York.

Pfurtscheller, G, Allison, BZ, et al. The hybrid BCI. *Frontiers in Neuroprosthetics,* 2:3. Published electronically April 10, 2010. doi: 10.3389/fnpro.2010.00003.

Prochazka, A, Gauthier, M, et al. (1997) The bionic glove: An electrical stimulator garment that provides controlled grasp and hand opening in quadriplegia. *Archives of Physical Medicine and Rehabilitation* 78:608–614.

Rosler, F. (2005) From single-channel recordings to brain-mapping devices: The impact of electroencephalography on experimental psychology. *History of Psychology* 8(1): 95–117.

Schalk, G. (2008) Brain-computer symbiosis. *Journal of Neural Engineering* 5(1): 1–15.

Stanslaski, S, Cong, P, et al. (2009) *An implantable bi-directional brain-machine interface system for chronic neuroprosthesis research.* Engineering in Medicine and Biology Society, Annual International Conference of the IEEE.

Stein, RB, Everaert, DG, et al. (2010) Long-term therapeutic and orthotic effects of a foot drop stimulator on walking performance in progressive and nonprogressive neurological disorders. *Neurorehabilitation and Neural Repair* 24(2): 152–67.

Chapter 4. Multitasking and the Metal Man: How Much Can Iron Man's Mind Manage?

Almor, A. (2008) Why does language interfere with vision-based tasks? *Experimental Psychology* 55(4): 260–68.

Coull, JT. (2004) fMRI studies of temporal attention: Allocating attention within, or towards, time. *Cognitive Brain Research* 21(2): 216–26.

Coull, JT, and Nobre, AC. (1998) Where and when to pay attention: The neural systems for directing attention to spatial locations and to time intervals as revealed by both PET and fMRI. *Journal of Neuroscience* 18: 7426–35.

De Jong, MJ. (2003) Cellular telephone use while driving: Growing awareness of the danger. *Journal of Emergency Nursing* 29(6): 578–81.

Friscolanti, M. (2006) *Friendly fire: The untold story of the U.S. bombing that killed four Canadian soldiers in Afghanistan.* John Wiley & Sons, New York.

Hunton, J, and Rose, JM. (2005) Cellular telephones and driving performance: The effects of attentional demands on motor vehicle crash risk. *Risk Analysis.* 25(4): 855–66.

Levin, A. (2010) FAA wants no pilot distractions. *USA Today,* April 25, 2010.

Nadkarni, NK, Zabjek, K, et al. (2010) Effect of working memory and spatial attention tasks on gait in healthy young and older adults. *Motor Control* 14(2): 195–210.

Redelmeier, DA, and Tibshirani, RJ. (1997) Association between cellular-telephone calls and motor vehicle collisions. *The New England Journal of Medicine* 336(7): 453–58.

Schmidt, RA, and Lee, TD. (2005) *Motor control and learning: A behavioral emphasis.* Human Kinetics, Champaign, Illinois.

Watson, JMS, and Strayer, DL. (2010) Supertaskers: Profiles in extraordinary multi-tasking ability. *Psychonomic Bulletin & Review* 17(4): 479–85.

Chapter 5. *Softening Up a Superhero: Why the Man with a Suit of Iron Could Get a Jelly Belly*

Hidler, J, Nichols, D, et al. (2009) Multicenter randomized clinical trial evaluating the effectiveness of the lokomat in sub-acute stroke. *Neurorehabilitation and Neural Repair* 23(1): 5–13.

Lam, T, Eng, JJ, et al. (2007) A systematic review of the efficacy of gait rehabilitation strategies for spinal cord injury. *Topics in Spinal Cord Injury Rehabilitation* 13(1): 32–57.

Lipnicki, D, and Gunga, H-C. (2009) Physical inactivity and cognitive functioning: results from bed rest studies. *European Journal of Applied Physiology* 105(1): 27–35.

Lipnicki, DM, Gunga, H-C, Belavý, DL, and Felsenberg, D. (2009) Bed rest and cognition: Effects on executive functioning and reaction time. *Aviation, Space, and Environmental Medicine* 80(12): 1018–24.

Moseley, AM, Stark, A, et al. (2008) Treadmill training and body weight support for walking after stroke. *Cochrane Database of Systematic Reviews.* 1.

Nash, MS, Jacobs, PL, et al. (2004) Metabolic and cardiac responses to robotic-assisted locomotion in motor-complete tetraplegia: A case report. *Journal of Spinal Cord Medicine* 27(1): 78–82.

Nicholas, SC, Doxey-Gasway, DD, et al. (1998) A link-segment model of upright human posture for analysis of head-trunk coordination. *Journal of Vestibular Research* 8: 187–200.

Paloski, W, Harm, DL, et al. (2000) *Postural changes following sensory reinterpretation as an analog to spaceflight.* Fourth European Symposium on Life Sciences Research in Space, 175–78.

Reschke, MF, Bloomberg, JJ, et al. (1998) Posture, locomotion, spatial orientation, and motion sickness as a function of space flight. *Brain Research: Brain Research Reviews.* 28(1–2): 102–17.

Stetz, MC, Thomas, ML, et al. (2007) Stress, mental health, and cognition: A brief review of relationships and countermeasures. *Aviation, Space, and Environmental Medicine* 78: B252–B260.

Williams, D, Kuipers, A, et al. (2009) Acclimation during space flight: Effects on human physiology. *Canadian Medical Association Journal* 180(13): 1317–23.

Zehr, EP, and Duysens, J. (2004) Regulation of arm and leg movement during human locomotion. *The Neuroscientist* 10(4): 347–61.

Zehr, EP, Hundza, SR, et al. (2009) The quadrupedal nature of human bipedal locomotion. *Exercise and Sport Sciences Reviews* 37(2): 102–8.

Chapter 6. Brain Drain: Will Tony's Gray Matter Give Way?

Cardinali, L, Frassinetti, F, et al. (2009) Tool-use induces morphological updating of the body schema. *Current Biology* 19(12): R478–R479.

Ehrsson, HH, Rosen, B, et al. (2008) Upper limb amputees can be induced to experience a rubber hand as their own. *Brain* 131(12): 3443–52.

Flor, H, Denke, C, et al. (2001) Effect of sensory discrimination training on cortical reorganisation and phantom limb pain. *Lancet* 357(9270): 1763–64.

Flor, H, Knost, B, et al. (2002) The role of operant conditioning in chronic pain: An experimental investigation. *Pain* 95(1–2): 111–18.

Follain, J. (2010) Facing the future. *The Sunday Times Magazine.* London, England, January 17, 2010, 40–45.

Ganguly, K, and Carmena, JM. (2009) Emergence of a stable cortical map for neuroprosthetic control. *Public Library of Science (Biology)* 7(7): e1000153.

Mercier, C, Reilly, KT, et al. (2006) Mapping phantom movement representations in the motor cortex of amputees. *Brain* 129(8): 2202–10.

Ramachandran, VS, and Altschuler, EL. (2009) The use of visual feedback, in particular Miller visual feedback, in fostering brain function. *Brain* 132: 1643–1710.

Sedwick, C. (2009) Practice makes perfect: Learning mind control of prosthetics. *PLoS Biology* 7(7): e1000152.

Suminski, A, Tkach, DC, et al. (2010) Incorporating feedback from multiple sensory modalities enhances brain-machine interface control. *Journal of Neuroscience* 30:16777–87.

Taub, E, Uswatte, G, et al. (2006) The learned nonuse phenomenon: Implications for rehabilitation. *Europa Medicophysica.* 42(3): 241–56.

Wolpaw, JR, and Carp, JS. (2006) Plasticity from muscle to brain. *Progress in Neurobiology* 78(3–5):233–63.

Chapter 7. Trials and Tribulations of the Tin Man: What Happens When the Human Machine Breaks Down

Brown, TG, Ouimet, MC, et al. (2009) From the brain to bad behaviour and back again: Neurocognitive and psychobiological mechanisms of driving while impaired by alcohol. *Drug and Alcohol Reviews* 28(4): 406–18.

Caldwell, JA, Mallis, MM, et al. (2009) Fatigue countermeasures in aviation. *Aviation, Space and Environmental Medicine* 80(1): 29–59.

Cernak, I, and Noble-Haeusslein, LJ. (2009) Traumatic brain injury: An overview of pathobiology with emphasis on military populations. *Journal of Cerebral Blood Flow and Metabolism* 30(2): 255–66.

Donelan, JM, Li, Q, et al. (2008) Biomechanical energy harvesting: Generating electricity during walking with minimal user effort. *Science* 319(5864): 807–10.

Eke-Okoro, ST. (1982) The H-reflex studied in the presence of alcohol, aspirin, caffeine, force and fatigue. *Electromyography and Clinical Neurophysiology* 22: 579–89.

Haqqani, HM, and Mond, HG. (2009) The implantable cardioverter-defibrillator lead: Principles, progress, and promises. *Pacing and Clinical Electrophysiology* 32(10): 1336–53.

Heidbüchel, H. (2007) Implantable cardioverter defibrillator therapy in athletes. *Cardiology Clinics* 25(3): 467–82.

Howland, J, Rohsenow, DJ, et al. (2011) The acute effects of caffeinated versus non-caffeinated alcoholic beverage on driving performance and attention/reaction time. *Addiction* 106:335–341.

Johnston, H. (2008) Knee brace harvests "negative work." Retrieved June 30, 2010, from http://physicsworld.com/cws/article/news/32812.

Kalmar, JM, and Cafarelli, E. (1999) Effects of caffeine on neuromuscular function. *Journal of Applied Physiology* 87(2): 801–8.

Kumar, S, Porcu, P, et al. (2009) The role of GABA(A) receptors in the acute and chronic effects of ethanol: A decade of progress. *Psychopharmacology (Berl)* 205(4): 529–64.

Li, Q, Naing, V, et al. (2009) Development of a biomechanical energy harvester. *Journal of Neuroengineering and Rehabilitation* 6: 22.

Madden, JD. (2007) Mobile robots: Motor challenges and materials solutions. *Science* 318(5853): 1094–97.

Mangels, A. (2008) *Iron Man: Beneath the armor.* Del Ray Books, New York.

Mathis, JT, and Clutter, JK. (2007) Evaluation of orientation and environmental factors on the blast hazards to bomb suit wearers. *Applied Ergonomics* 38(5): 567–79.

Maughan, R. (2002) The athlete's diet: Nutritional goals and dietary strategies. *Proceedings of the Nutrition Society* 61(1): 87–96.

Melgaard, B, Saelan, H, et al. (1986) Symptoms and signs of polyneuropathy and their relation to alcohol intake in a normal male population. *Acta Neurologica Scandinavica* 73(5): 458–60.

Paradiso, JA, and Starner, T. (2005) Energy scavenging for mobile and wireless electronics. *Pervasive Computing* Jan–Mar: 18–27.

Reissig, CJ, Strain, EC, et al. (2009) Caffeinated energy drinks: A growing problem. *Drug and Alcohol Dependence* 99(1–3): 1–10.

Ritenour, AEMD, and Baskin, TWMD. (2008) Primary blast injury: Update on diagnosis and treatment. *Critical Care Medicine* 36:S311–S317.

Schweizer, TA, and Vogel-Sprott, M. (2008) Alcohol-impaired speed and accuracy of cognitive functions: A review of acute tolerance and recovery of cognitive performance. *Experimental and Clinical Psychopharmacology* 16(3): 240–50.

Stuhmiller, JH. (1997) Biological response to blast overpressure: A summary of modeling. *Toxicology* 121(1): 91–103.

Sullivan, EV, Harding, AJ, et al. (2003) Disruption of frontocerebellar circuitry and function in alcoholism. *Alcohol Clinical and Experimental Research* 27(2): 301–9.

Sullivan, E, and Pfefferbaum, A. (2005) Neurocircuitry in alcoholism: A substrate of disruption and repair. *Psychopharmacology* 180(4): 583–94.

Tarnopolsky, MA. (1994) Caffeine and endurance performance. *Sports Medicine* 18: 109–25.

Tarnopolsky, MA. (2008) Effect of caffeine on the neuromuscular system: Potential as an ergogenic aid. *Applied Physiology Nutrition and Metabolism* 33(6): 1284–89.

Trusty, JM, Beinborn, DS, et al. (2004) Dysrhythmias and the athlete. *AACN Advanced Critical Care* 15(3): 432–48.

Wallner, M, and Olsen, RW. (2008) Physiology and pharmacology of alcohol: The imidazobenzodiazepine alcohol antagonist site on subtypes of GABAA receptors as an opportunity for drug development? *British Journal of Pharmacology* 154(2): 288–98.

Chapter 8. Visions of the Vitruvian Man: Is Invention Really Only One Part Inspiration?

Abrams, M. (2006) *Birdmen, batmen, and skyflyers: Wingsuits and the pioneers who flew in them, fell in them, and perfected them.* Three Rivers Press, New York.

Bunch, BH, Hellemans, A. (2004) *The history of science and technology: A browser's guide to great discoveries, inventions, and the people who made them from the dawn of time to today.* Houghton Mifflin Company, New York.

Cajal, SRY. (1989) *Recollections of my life.* The MIT Press, Cambridge, Massachusetts.

Darwin, FE. (1958) *The autobiography of Charles Darwin and selected letters.* Dover, New York.

Gelb, MJ. (2004) *How to think like Leonardo da Vinci: Seven steps to genius every day*. Delta Books, New York.

Gifford, C. (2009) *10 inventors who changed the world*. Kingfisher, New York.

Glenn, J. (1996) *Scientific genius: The twenty greatest minds*. Saraband, Inc., Rowayton, Connecticut.

Kuhn, T. (1996) *The structure of scientific revolutions*. University of Chicago Press, Chicago.

McNichol, T. (2006) *AC/DC: The savage tale of the first standards war*. Jossey-Bass, San Francisco, California.

Poincare, H. (1921) *The foundations of science*. New York, Science Press.

Schwartz, EI. (2002) *The last lone inventor: A tale of genius, deceit, and the birth of television*. Perennial, New York.

Simonton, D. (1988) *Scientific genius*. New York, Cambridge University Press.

Chapter 9. Deal or No Deal? Could Iron Man Exist?

Allison, BZ, Wolpaw, EW, et al. (2007) Brain-computer interface systems: Progress and prospects. *Expert Review of Medical Devices* 4(4): 463–74.

Barbeau, H. (2003) Locomotor training in neurorehabilitation: Emerging rehabilitation concepts. *Neurorehabilitation and Neural Repair* 17(1): 3–11.

Behrman, AL, Bowden, MG, et al. (2006) Neuroplasticity after spinal cord injury and training: An emerging paradigm shift in rehabilitation and walking recovery. *Physical Therapy* 86(10): 1406–25.

Behrman, AL, and Harkema, SJ. (2007) Physical rehabilitation as an agent for recovery after spinal cord injury. *Physical Medicine and Rehabilitation Clinics of North America* 18(2): 183–202.

Cernak, I, and Noble-Haeusslein, LJ. (2010) Traumatic brain injury: An overview of pathobiology with emphasis on military populations. *Journal of Cerebral Blood Flow and Metabolism* 30(2): 255–66.

Christie, BR, Eadie, BD, et al. (2008) Exercising our brains: How physical activity impacts synaptic plasticity in the dentate gyrus. *Neuromolecular Medicine* 10(2): 47–58.

Cyberdyne. (2010) www.cyberdyne.jp/english/robotsuithal/index.html

Daly, JJ, and Wolpaw, JR. (2008) Brain-computer interfaces in neurological rehabilitation. *The Lancet Neurology* 7(11): 1032–43.

Dobkin, BH. (2008) Training and exercise to drive poststroke recovery. *Nature Clinical Practice Neurology* 4(2): 76–85.

Ganguly, K, and Carmena, JM. (2009) Emergence of a stable cortical map for neuroprosthetic control. *PLoS Biology* 7(7): e1000153.

Jones, RN, Fong, TG, et al. (2010) Aging, brain disease, and reserve: Implications for delirium. *American Journal of Geriatric Psychiatry* 18: 117–27.

Kallus, KW, Hoffmann, P et al. (2010) The taskload-efficiency-safety-buffer triangle: Development and validation with air traffic management. *Ergonomics* 53(2): 240–46.

Leach, JB, Achyuta, AKH, et al. (2010) Bridging the divide between neuroprosthetic design, tissue engineering and neurobiology. *Frontiers in Neuroengineering* 8: 2–18.

Mehrholz, J, Kugler, J, et al. (2008) Locomotor training for walking after spinal
cord injury. *Cochrane Database of Systematic Reviews* 1.

MIT. (2010). Inventor of the week: Yoshiyuki Sankai. Retrieved February 6,
2010, from http://web.mit.edu/invent/iow/sankai.html.

Moseley, AM, Stark, A, et al. (2008) Treadmill training and body weight support for walking after stroke. *Cochrane Database of Systematic Reviews* 1.

Nithianantharajah, J, and Hannan, AJ. (2009) The neurobiology of brain and
cognitive reserve: Mental and physical activity as modulators of brain
disorders. *Progress in Neurobiology* 89(4): 369–82.

Nuytco. (2010) About Nuytco—Phil Nuytten. Retrieved February 6, 2010, from
www.nuytco.com/about/phil.shtml.

Ritenour, AEMD, and Baskin, TWMD. (2008) Primary blast injury: Update on
diagnosis and treatment. *Critical Care Medicine* 36:S311–S317.

Schalk, G. (2008) Brain-computer symbiosis. *Journal of Neural Engineering* 5(1):
1–15.

Stern, Y. (2009) Cognitive reserve. *Neuropsychologica* 47(10): 2015–28.

Wang, W, Collinger, JL, et al. (2010) Neural interface technology for rehabilitation: Exploiting and promoting neuroplasticity. *Physical Medicine and
Rehabilitation Clinics in North America* 21(1): 157–78.

Wolpaw, JR, Birbaumer, N, et al. (2002) Brain-computer interfaces for communication and control. *Clinical Neurophysiology* 113(6): 767–91.

Wolpaw, JR, and Carp, JS. (2006) Plasticity from muscle to brain. *Progress in
Neurobiology.*

Wolpaw, JR, and McFarland, DJ. (2004) Control of a two-dimensional movement signal by a noninvasive brain-computer interface in humans. *Proceedings of the National Academy of Science (US)* 101(51): 17849–54.

Zeitler, DM, Budenz, CL, et al. (2009) Revision cochlear implantation. *Current
Opinion in Otolaryngology & Head and Neck Surgery* 17(5): 334–38.

Index

Page numbers in *italics* indicate figures.

Abrams, Michael, *Birdmen, Batmen, & Skyflyers,* 136
acetylcholine, 24
action potentials, 54
activation of muscles, 21
Afghanistan, Tarnak farm incident in, 122–23
aftereffects, 85, 105–9
aging: biological, 176–78; multitasking and, 72–73
Albinus, Bernhard Siegfried, 6
alcohol: caffeine in drinks with, 122; effects of on body, 116–19; use and abuse of by Stark, 101–2, 116, 119–21, 157
ALS. *See* amyotrophic lateral sclerosis
Altholz, Addison, 132
amputation of limbs, 98–99, *100,* 101–2
amyotrophic lateral sclerosis (ALS), 56, 58
Anglesea leg, 27
ankle, flexing of when walking, 40–42, *41*
anthropomorphic suits, 19
apoptosis, 117
Argo Medical Technologies, 87
armor: adaptation to, 91, *92*; classic red and gold, *7,* 9–11, 144–45, *145, 146*; computer malfunctions and, 128–30; concept of modular, 8, *9,* 10; deep water suit, 142, *143*; as defining characteristic for Iron Man, 4; effects of wearing, 81–82, 85–86, 93, 102, 105–9; functions of, 5–6; Golden Avenger, 8–9; Hulk-buster, 6, 11; need for simplification of, 70–71; original gray, *7,* 8–9, *9*; for protection, 123–24, *125,* 126, 171; reactions to, 80–81; to restore lost function, 90; sensors on, 107; stealing and use of, 155–56; time needed to develop, 65–66, 144, 147; training needed to operate, 155, 159, 161–62; types of, 6, *7,* 8; views of, 6; War Machine, 77–79, *78,* 152, 156–57. *See also* brain-computer interface; brain-machine interface; Extremis armor; Neuromimetic Telepresence Unit (NTU-150) armor
Armor Wars story arc, 155–56
arousal level and stress, 76–77, *77*
artificial intelligence, 129
Asimov, Isaac: *I, Robot,* 144; "Run-around," 129–30
"The Assassination of the Nyctalope" (Hire), 5
assistive devices, 87
assistive technology: adaptation to, 91, *92,* 93; to amplify performance, 90
astronauts: deconditioning effects of space on, 83–85, *84,* 86; suits worn by, 109, 110
atrial fibrillation, 127

attention, limits of, 70–75
Avatar (movie), 13, 156
axons, *23*, 117

Bagley, Marc, 120
Baskin, Toney, 124
Batman: exoskeleton of, 4; Iron Man
    compared with, xii; length of career of,
    178–79; as "possible" superhero, 120;
    Bruce Wayne and, 119. See also
    Becoming Batman
batteries, energy storage in, xiv
BCI. *See* brain-computer interface
BDNF (brain-derived neurotrophic
    factor), 176
Becoming Batman (Zehr): description of,
    xii; issues introduced in, 82, 154; length
    of career of Batman, 178–79; martial
    arts training, 164; writing of, 150–51
bed rest, 82
Bendis, Brian Michael, 164
Berger, Hans, 55
Bikila, Abebe, 30–31
bionic glove, 42
bionics, definition of, 4
Birdmen, Batmen, & Skyflyers (Abrams),
    136
Black Sabbath and "Iron Man" song, 3–4
blast waves from explosions, 124, *125*,
    171–72
body: adaptability of, 94–95; effects of
    gravity on, 84, 85–86; effects of
    stimulants on, 121–23; effects of stress
    on, 82; retraining of, 87–88, *89*, 90–91;
    temperature of when wearing suit of
    armor, 81; weight of muscles in, 17.
    See also movement, biological; *specific
    parts of body*
body-weight assisted treadmill training, 87
Boldrey, Edwin, 52
Bolt, Usain, 36
bomb disposal suits, 124, *125*, 171
bone density with reduced activity level,
    83, 84, 86
brain: cerebellum, 117–19, *118*; complex-
    ity of, 51; concussions to, 166–67,
    168, 169–72; cortex of, 50, 51–53, *53*;
    deep brain stimulation, 62; electrical

signaling of, 49, 55; electrical stimula-
    tion of, 52; electrodes implanted in,
    160–61, 174, *175*; frontal cortex, 167,
    168; increasing representation of body
    in, 102–3, *104*, 105; motor cortex of, 21,
    52, 54; neurons in, 51, 52; occipital
    cortex, 167, *168*; plasticity of, 90, 96,
    109, 170; somatosensory cortex, 95–96;
    specialized areas and functions of, 48,
    49, 51, 55. *See also* brain-computer
    interface; brain-machine interface;
    frontal lobes
brain-computer interface (BCI): ability
    to use armor and, 157; changes in with
    use of prosthetic limb, 102–3, *104*, 105;
    EEG and, 55; science of, 47–51; as skill
    development, 159–60
brain-derived neurotrophic factor
    (BDNF), 176
BrainGate, 59–61, *60*
brain-machine interface: ability to use
    armor and, 157; for ALS, 56, 58; in
    Avatar, 156; description of, 58–61;
    jacking into satellite, 96; multitasking
    in, 161; physical and mental condition
    for, 158–62; potential for use in
    rehabilitation, *158*; sensation from
    robotic limbs and, 107–8; in telepres-
    ence armor, 12–13, 61–62, *63*, 64;
    training to use, 155, 159–60, 161–62
brain reserve, 178
Brevoort, Tom, 14–15
Broca, Paul, 51
Burkett, Brendan, 29–30
Buschman, Timothy, 51

caffeine, as stimulant, 121–22
Cajal, Santiago Ramon y, 149–50
calculus, 132
Caldwell, J. Lynn, 122
Captain America, 164, *165*
Caramagna, Joe, 155
Card, Orson Scott, *Ultimate Iron Man*,
    101, 120
Cardinali, Lucilla, 108, 109
cardiovascular system, with reduced
    activity level, 82, *83*. *See also* heart
Carmena, Jose, 102, 103, 105

Carson, Richard, 99
Caton, Richard, 55
cell bodies, *23*
cell phones, multitasking with, 71–72, 76
cells: death of, 117; ions in, 20; pacemaker, 126. *See also* neurons
cellular pumps (Na-K ATPase), 20
center of mass (COM), 31, 33, *34*
central pattern generators, 90–91
cerebellum, damages to in alcoholism, 117–19, *118*
Cernak, Ibolja, 171
cervical region of spine and spinal cord, 21, *22*, 58–59
chest plate, xiii, xiv, 128. *See also* implantable cardioverter defibrillator
chloride, 20
classic red and gold armor: description of, *7*, 9–11; HAL suit compared with, 144–45, *145*, *146*
closed loop control, 62
cochlear prosthetics/implants, 42, 173, 174, 176
cognitive load, 70
cognitive reserve, 178
Colossus, 4
COM (center of mass), 31, 33, *34*
combat aviation, as art, 123
common peroneal nerve, 40
computer malfunctions, 128–30
concussion, risks of, 166–67, *168*, 169–72
contractile proteins, 24
contractions, shortening, 26
contre coup in concussions, 167
cortex of brain, role of in movement, 50, 51–53, *53*
creativity and genius: description of, 132; force and, 150–51; formal education and, 149; open minds for, 133–34; in scientific discovery, 148; targeting, 133
Crosby, Sidney, 169
Curie, Marie, 149
Cyberdyne Inc., 43, 144, 147. *See also* HAL (Hybrid Assistive Limb) robot suit
CyberKinetics BrainGate, 59–61, *60*
Cybermen (*Dr. Who* characters), 4–5
cybernetics, definition of, 4
cyborgs, 4–5, 14

Dar, Kathleen (character), 87
The Dark Knight Returns (Miller), 4
Darwin, Charles, 149, 150
da Vinci, Leonardo, 132, 133
deconditioning effects: on astronauts, 83–85, *84*, 86; of wearing armor, 82–83, *83*, 85–86
deep brain stimulation, 62
deep diving and underwater exosuits, 141–42, *143*
defibrillator, implantable cardioverter (ICD). *See* implantable cardioverter defibrillator
De Jong, Marla, 71
Demon in a Bottle story arc, 119–21
depressants. *See* alcohol
dermatomes and nerves in spine, 21, *22*
Descartes, René, 50
Devauchelle, Bernard, 97
Dinoire, Isabelle, 97, 98
Downey, Robert, Jr., xiii, 165
dual task paradigm, 72–75
Dubernard, Jean-Michel, 97
du Petit, Francois Pourfour, 50

eccentric contractions, 26
Edison, Thomas, 148
EEG. *See* electroencephalography
Ehrsson, Henrik, 106
Einstein, Albert, 131–32, 148–49
electrical signaling of brain, 49, 55
electrical stimulation: of brain, 52; of neurons, 55–56
electrodes implanted in brain, 160–61, 174, *175*
electroencephalography (EEG), 55–56, *57*, 95
electromyography (EMG), 58
Ellis, Warren: *Extremis*, 109, 146; Extremis armor, 14
embodiment: armored exoskeletons and, 109–10; Extremis armor and, 14; pilots and, 76; prosthetic limbs and, 106–7, *107*; tools and devices and, 108–9
Emotiv, 58
encapsulation, process of, 174, *175*
energy drinks, caffeinated, 122
epilepsy, 51, 160–61

ergogenic aids, 121
ethanol. *See* alcohol
excitable tissue, 19–20, 49
exercise: to counter deconditioning
   effects, 86; in training to use armor,
   177–78
exoskeleton: for ankle joint, 33, *34*; of
   Batman, 4; embodiment and, 109–10;
   Lokomat robotic, 87–88, *89,* 90, 91;
   lower limb, *35,* 36; neurorobotics,
   87–88, *89,* 90; robotic, structure and
   function of human limbs compared
   with, 19–20; underwater, 141–42, *143*
Exosuit, 142, *143*
Extremis armor: as best possible, 112,
   179–80; depiction of, *7*; description of,
   13–15; origin story, 147; Tony Stark on,
   134

face shield, 10, *11,* 171
face transplants, 97–98
fast twitch motor units, 25
Favreau, Jon, 3
Fernandez, Eduardo, 62
Ferris, Dan, 33
FES. *See* functional electrical stimulation
fibula, 29, 40
fingers, neuroprosthetics for, *44,* 45
fixed wing flyers, 135–36, *137,* 138–41
flying, "events" when, 76
fMRI. *See* functional magnetic resonance
   imaging
Follain, John, 98
foot drop, 40
force production in muscles: changing,
   24; oddities in, 25–26
force-velocity relation, 25–26
Fraction, Matt, 96
Freud, Sigmund, 149
friendly fire accident in Afghanistan,
   122–23
Friscolanti, Michael, *Friendly Fire,* 123
Fritsch, Gustav, 51–52
frontal cortex, 167, *168*
frontal lobes: alcoholism and, 119;
   concussions and, 169; functions of, *49,* 51
functional electrical stimulation (FES),
   39–40, *41,* 42

functional magnetic resonance imaging
   (fMRI), 96

GABA_a receptors, 116
Galen, 49–50
Galvani, Luigi, 50
Ganguly, Karunesh, 102, 103, 105
genius: creativity and, 132, 133–34, 148,
   149, 150–51; Tony Stark as, 147–48, 151.
   See also invention and genius
Golden Avenger armor, 8–9
Golgi, Camillo, 150
"go pills," 121–23
Granov, Adi, 14, 146
gravity, effects of on body, 84, 85–86
Gross, Charles, 50

HAL (Hybrid Assistive Limb) robot suit,
   43, 144–47, *145, 146*
Hammer, Justin (character), 130,
   153, 179
hands: neuroprosthetics for, 43, *44,* 45;
   Prehensor, 142, *143*
hands-free phones, 72, 75
Hansen, Maya (character), 14, 146–47
Hatsopoulos, Nicholas, 107
head injuries: from blast waves, 171–72;
   concussions, 166–67, *168,* 169–72;
   motor control problems after, 48–49
headset, user interface, 47, 62
heart: arrhythmia of, 126–28; atria of,
   126–27, *128*; functions of, 126–27;
   implantable defibrillator for, 127–28,
   128; origin story of shrapnel lodged in,
   126; ventricles of, 126–27, *128. See also*
   chest plate; implantable cardioverter
   defibrillator
Heck, Don, xii, 182
hip hiking, 40
hippocampus, 169
Hire, Jean de la, 5
Hitzig, Eduard, 51, 52
Hocama, Inc., 87
homunculus concept, 52, *53*
Howland, Jonathan, 122
Hulkbuster armor, 6
human flight, 135–36, *137,* 138–41
Hunton, James, 73

Hybrid Assistive Limb (HAL) robot suit.
    See HAL (Hybrid Assistive Limb) robot
    suit
hydraulic actuators, 65

ICD. See implantable cardioverter
    defibrillator
I, Cyborg (Warwick), 38
IEDs (improvised explosive devices), 124
i-LIMB hand, 43, 44, 45
Illuminati, 164
immune response to implants, 174, 175
implantable cardioverter defibrillator
    (ICD), 127–28, 128, 173–74, 175, 176
improvised explosive devices (IEDs), 124
inattention blindness, 71–72
inertial aspects of body, 86
Infantino, Carmine, 119
information processing, limits of, 70–75
injuries from bomb blasts, 124, 126,
    171–72
innervation ratio, 176–77
insights and force, 150–51
International Amateur Athletic
    Association, 29
International Collaboration on Repair
    Discoveries, 88
invention and genius: description of,
    131–32; open minds for, 133–34; patent
    law and, 151–53; personality for, 140,
    147; relevant to robotic armor, 134–35;
    Tony Stark and, 147–51; targeting, 133,
    140
inventor, definition of, 152
Invincible Iron Man comics: #1 (2008),
    115, 183; #2 (2008), 69; #8 (2009), 97;
    #9 (2009), 37, 154; #10 (2006), 154; #14
    (2009), 110; #47 (1972), 16, 91, 92, 94;
    #70 (1974), 179; #129 (1979), 16; #169
    (1983), 156–57; #170 (1983), 156–57,
    161–62; #280 (1992), 12, 13; #281
    (1992), 156; #282 (1992), 156; #284
    (1992), 78, 156, 182; #290 (1993), 12, 13,
    46, 47, 101, 182; #291 (1993), 47; Demon
    in a Bottle story arc, 119–21
I, Robot (Asimov), 144
Iron Man: Batman compared with, xii;
    characteristics of, xi; length of career

of, 178–79; as metaphor, x; origin story,
    xiii, 14, 126, 182; possibility of becom-
    ing, 155–57, 158; as "possible" superhero,
    166; project development phases, 65–66,
    111–12; reinvention of, ix–x; as without
    superpowers, xiv–xv. See also armor;
    Stark, Tony
Iron Man (movie): action figures, 10, 11;
    concussive events in, 171; description
    of, 80, 183; Stark on pilots, 163
Iron Man: Beneath the Armor (Mangels),
    xiv, 3, 120
Iron Man / Captain America: Casualties
    of War #1 (2007), 115
Iron Man: Civil War (graphic novel), 148,
    163–64, 165
Iron Man comics: #1–6 (2005–2006), 183;
    #13 (2009), 70–71; #142 (1980), 94;
    #144 (1980), 80; #182 (1984), 120; #200
    (1985), 182; #218 (1987), 142, 143; #242
    (1989), 87; #244 (1989), 90; #245 (1989),
    90; volumes of, 181
Iron Man: Director of S.H.I.E.L.D.: #31
    (2007), 80, 96, 133–34; #33 (2007), 96
Iron Man: Extremis (Ellis), 109, 131,
    146–47
Iron Man: Hypervelocity: #1–6 (2007),
    183; #2 (2007), 37; #5 (2007), 69;
    graphic novel, 170; story arc, 129
Iron Man: The End (graphic novel), 166,
    170
Iron Man: The Official Index to the Marvel
    Universe (Marvel Comics), 181
Iron Man 2 (movie): action figures, 78;
    alcohol abuse shown in, 116; armor in,
    79, 107, 151–52, 156–57; description of,
    183; fighting in, 163; Jim Rhodes in, 74;
    safeguards for control systems in, 130;
    Whiplash in, 179
Iron Monger, 171
Iron Monger armor, 6, 7, 8
Irons, John Henry (character), 4
isometric contractions, 26

Jackson, John Hughlings, 51
jet fighter pilot training, 163
jet-pack flight, 136, 137, 138–41
Jevons, William, 148

Kakalios, Jim, xiv
Kennedy, Sal (character), 147
Kirby, Jack, xii, 8
Kubert, Andy, 120
Kuhn, Thomas, *The Structure of Scientific Revolutions*, 149

Lam, Tania, 88
Layton, Bob, 119
Leach, Jennie, 174
Lee, Stan, xii, xiv, 3, 94, 182
legs, prosthetic, 27, 29
Lieber, Larry, xii, 182
Liebnitz, Gottfried, 132
limbs: amputation of, 98–99, *100, 101*–2; changes in BCI with use of prosthetics, 102–3, *104*, 105; lower, exoskeleton for, *35*, 36; nervous system commands to, 39, 45, *46*, 47; prosthetic, 27, 29; robotic, sensation from, and brain-machine interface, 107–8; structure and function of compared with exoskeleton, 19–20
Lokomat, 87–88, *89, 90*, 91
long-term potentiation (LTP), 117
Lord of the Rings trilogy, 81–82
Lou Gehrig's disease (ALS), 56, 68
lumbar region of spine and spinal cord, 21, *22*

Magma, 157
Mangels, Andy: *Iron Man: Beneath the Armor*, xiv, 3, 120; on serum, 14
man-machine hybrids, 4–5
martial arts: empty hand training, 108; Tony Stark's training in, 163–65, *165*
Marvel Comics: *Iron Man: The Official Index to the Marvel Universe*, 181; Secret Invasion story arc, 129. *See also* Invincible Iron Man comics; Iron Man comics
"Masters of Silence," 156
Mattel Mind Flex, 58
Medtronic Neuromodulation Technology Research division, 62
membrane potential, 20–21
Michelinie, David, xiv–xv, 119, 120
microglial activation, 174, *175*

military aspect of caffeine ingestion, 122–23
military aviation, multitasking in, 75–79
Miller, Earl, 51
Miller, Frank, *The Dark Knight Returns*, 4
Monroe, Alexander, 50
Monty Python's Flying Circus (TV series), 133
motoneurons: aging and, 176–77; description of, 21, *23*, 23–24
motor cortex of brain, 21, 52, 54
motor evoked potentials, 56
motor maps, 95, 108
motor units, 23–25
movement, biological: after injuries to brain, 48–49; cerebellum and, 117–19, *118*; chain of commands in, 17; cortex and, 50, 51–53, *53*; effects of alcohol on, 121; effects of caffeine on, 122; motor units, 21, *23*, 23–24; multiple tasks and, 70–75; muscles in, 16–17, 19–20; planning and coordination of, 54–56, *57*, 58; robotic assistance with, 91; sensation from, 97; signals for, 48, 53–54; timescales for, 162; walking and running, 31–32, *32*
movement, machine-based, 36
multitasking: in brain-machine interface, 161; problems with, 70–75; training in, 75–79
muscle cells of heart, 126
muscle fibers, *23*, 23–24, 176–77
muscles: in biological movement, 16–17, 19–20; in body weight, 17; contractions during movement, 26; force of gravity and, 85–86; major, 17, *18*; monitoring to make motors move, 39–45; as motors, 21, *22*, 23–26; underuse of, 82–83, *83*
myelin, 24

Na-K ATPase (cellular pump), 20
nanotechnology, 112
nerves: monitoring to make motors move, 39–45; in spine and dermatomes, 21, *22*
nervous system: as adaptable and changeable, 90; commands to limbs from, 39, 45, *46*, 47; damage to from

blast waves, 172; depicted in *Ultimate Iron Man* graphic novel, 120–21; effects of alcohol on, 116–19; electrical nature of, 50; excitable tissue in, 19–20, 49; with reduced activity level, *83*; stimulants for, 121–23; working against normal decline of, 176–78
neural plasticity/neuroplasticity, 90, 96, 109, 170
Neuromimetic Telepresence Unit (NTU-150) armor: brain-machine interface and, 61–62, *63*, 64; depictions of, *7, 46*; description of, 11–13; sensation feeding back into, 62
neuromuscular junction, 24
neuronal recordings, 111–12
neuron doctrine, 150
neurons: in brain, 48, 51, 52; central pattern generators, 90–91; electrical stimulation of, 55–56; membrane potential, 20–21; motor, 21, *23,* 23–24, 176–77; as tuned to movement directions, 103; upper motor, 54, 56
neuropathy from alcoholism, 117
neuroprosthetics: bionic glove, 42; cochlear, 42, 173, 174, 176; definition of, 38; general principles for, 38–39; HAL robot suit, 43, 144–47, *145, 146*; i-LIMB hand, 43, *44,* 45; immune response to, 174, *175*; improvements in, 105; optical, 62, *63,* 64; Pro-Digits, *44,* 45; types of, 42; WalkAide, 40–42, *42.* *See also* prosthetics
neurorobotics, 87–88, *89,* 90
NeuroSky, 58
neurotransmitters and alcohol, 116–17
"neuro-web life support system," 12
New Avengers: Illuminati comic miniseries (2007), 164
Newton, Isaac, 132
Newtsuit, 141–42, *143*
Nichols, David C., 123
NMDA receptors, 116–17
Noble-Haeusslein, Linda, 171
NTU-150 armor. *See* Neuromimetic Telepresence Unit armor
nucleus, *23*

Nuytten, Phil, 134, 141–42, *143*
Nyctalope (cyborg), 5

occipital cortex, 167, *168*
occipital lobe, *49*
O'Neil, Dennis, 119, 120
open loop control, 62
optical neuroprosthesis, 62, *63,* 64
Ossur Flex-Foot Cheetah prosthetic, 29, *30*

Palmiero-Winters, Amy, 36
paradigm shifts, 149
parietal lobe, *49*
Parkinson's disease, 62
patent law, 151–53
Penfield, Wilder, 52
perception of time, and attention, 75
performance and stress, 76–77, *77*
perturbations, 85
phantom limb and phantom pain syndromes, 98–99, *100,* 101–2
The Physics of Superheroes (Kakalios), xiv
physiological adaptation to stress, effects of on body, 82
pilots: multitasking by, 73, 75–79; training of, 163
Pistorius, Oscar, 29, *30,* 31, 33
Planck, Max, 149
planning of movement, 54–56, *57,* 58
Poincaré, Henri, 150
post-concussion syndrome, 170
potassium, 20
Potts, Pepper (character), 110, 152
powered devices, state of technology in, 26–27
Prehensor, 142, *143*
premotor area of brain, 55
pressure waves from explosions, 124, *125,* 171–72
Prochazka, Arthur, 42
Pro-Digits, *44,* 45
Project Cyborg, 38
prosthetic motor memory, 105
prosthetics: changes in BCI with use of, 102–3, *104,* 105; early examples of, 27, *28*; embodiment and, 106–7, *107*; including sensation from, 107–8;

prosthetics *(cont.)*
  performance enhancement due to, 33,
    35–36; of Pistorius, 29, *30*, 31, 33. *See*
    also neuroprosthetics
  protection, use of armor for, 123–24, *125*,
    126, 171
proteins in muscle fibers, 24

Quesada, Joe, ix

Ramachandran, Vilayanur, 99
reactive gliosis, 174, *175*
recruitment, 24
Red Barbarian, 156
Reed, Brian, 164
Reeve, Christopher, 58–59
rehabilitation: potential for use of
    brain-machine interface in, *158*; robots
    for, 86–88, *89*, 90; robot suit for, 144–47,
    145, 146; of walking, 90–91
reinnervation, 97–98, 177
Reissig, Chad, 122
remote control armor system, 12–13
retraining of body, 87–88, *89*, 90–91
ReWalk device, 87, 162
Rhodes, Jim "War Machine" (character):
    armor given to Air Force by, 152; battle
    with, 116; first use of armor, 78–79,
    156–57; in *Iron Man* movie, 163; in *Iron*
    Man 2 movie, 130; on jacking into
    satellite, 96; on "Masters of Silence,"
    156; multitasking by, 74, *74*; practice
    using armor, 161–62; stimulants and,
    122–23
risks of being Iron Man: to implants,
    173–74, *175*, 176; to nervous system,
    176–78; overview of, 166; whiplash,
    166–67, *168*, 169–72
risks taken by inventors, 136, 138
Ritenour, Amber, 124
robotic devices: direct connections
    between nervous system and, 45, *46*,
    47; HAL suit, 43, 144–47, *145*;
    Prehensor, 142, *143*
robot Iron Man. *See* Neuromimetic
    Telepresence Unit (NTU-150) armor
Romita, John, Jr., 119
Rose, Jacob, 73

Rossy, Yves, 134, 135–36, *137*, 138–41
Rourke, Mickey, 74
Rousseau, Craig, 155
rubber hand illusion, 106, *107*
"Runaround" (Asimov), 129–30
running, 31–33, *32, 34*, 35–36. *See also*
    walking

sacral region of spine and spinal cord, 21,
    22
safeguards for control systems of armor,
    129–30
Sankai, Yoshiyuki, 134, 144–47
satellite, jacking into, 96
scientific advance and discovery, 147–51
Scientific Genius (Simonton), 148, 149, 150
secondary impact syndrome, 170
Secret Invasion (graphic novel), 129
senescence, 176–78
sensors on armor, 107
sensory adaptation, 94–95
sensory areas of brain, 52
sensory input in human flight, 139
sensory substitution, 101
shell shock syndrome, 124, 126
Sherrington, Charles, 24
Shippam-Brett, Cynthia, 152
Simek, Art, 182
Simonton, Dean, *Scientific Genius*, 148,
    149, 150
The Six Million Dollar Man (TV series), 4
skeletal muscle, 17
Skrull, 129, 164
skyflying, 135–36, *137*, 138–41
slow twitch motor units, 25
sodium, 20
sodium channels in axons, 117
software, problems with, 129
Sohn, Clem, 138
somatosensory cortex, 95–96
somatosensory evoked potentials, 56
somatosensory maps: changes in, 97–98;
    description of, 95–96; phantom limb
    and phantom pain syndromes, 98–99,
    100, 101–2; tools and devices and,
    108–9
speech, attention demands of, 71–72
Spiderman, 115–16

spinal cord: action potentials, 54; in ALS, 56; description of, 21, 22; injuries to, 58–59, 88; neurons in, 48, 90–91
sports and concussions, 166–67, 169
sports biomechanics, 27
"spot checks" by police for intoxication, 118
springlike effect while walking and running, 31–33, 34
sprouting, 177
Stableford, Brian, 5
Stane, Obidiah (character), 6, 7, 8, 77, 77
Stanslaski, S., 62
Stark, Howard (Tony's father), 179
Stark, Tony (Anthony Edward, character): alcohol use and abuse by, 101–2, 116, 119–21, 157; chest plate, xiii, xiv, 128; as genius inventor, 147–48, 151, 152; health crises of, 87; height and weight of, 17; Iron Man origin story, xiii, 14, 126, 182; as ladies' man, 8; nervous system degeneration, 101; on pilots, 163; reinvention of, ix–x. See also Iron Man; risks of being Iron Man
Star Trek: the Next Generation (TV series), 64
steam pumps, 25
Stein, Richard, 41
stimulants, effects of on body, 121–23
Strayer, David, 74–75
strength training, 177
stress: effects of on body, 82; performance and, 76–77, 77
stretch reflexes, 31, 32
The Structure of Scientific Revolutions (Kuhn), 149
Suminski, Aaron, 107
superheroes: as superhuman, 120; viewed as "possible," 154–55. See also specific characters
Superman, 115–16
supertaskers, 75
supplementary motor area of brain, 55
surfaces for walking or running, 32–33
Surrogates (movie), 13
synaesthesia, 101
synapses, 23, 24

Tales of Suspense: #39 (1963), xii, 8, 9, 91, 128, 182; #40 (1963), xiii–xiv, 9; #48 (1963), 9–10, 182
Tarnak farm incident in Afghanistan, 122–23
Taylor, Charlotte, 50
telepresence armor. See Neuromimetic Telepresence Unit (NTU-150) armor
temporal lobe, 49
Terminator, 4
texting while driving, 72
therapeutic electrical stimulation, 39
thermoregulation, 81
thoracic region of spine and spinal cord, 21, 22
Thorbuster armor, 11
thoughts, transforming into actions of machines, 37–38
Three Laws of Robotics (Asimov), 129–30
tibia, 29
tools, use of, 108–9
Touch Bionics, 43, 44
training to use armor: exercise, 177–78; level of, and multitasking, 73, 75–79; time needed, 155, 159–60, 161–62; type needed, 162–65
transcranial magnetic stimulation, 55–56, 57
tremor, physiological, 25

Ultimate Iron Man (Card), 101, 120
Ultimate Iron Man (vol. 1, 2006), 131
upper motor neurons, 54, 56
Urban, Karl, 81–82

Valentin, Leo, 138
Vanko, Ivan "Whiplash" (character), 74, 130, 179
video game controllers, 58
virtual environments, human interaction with, 13
vision, as primary sense, 101, 106
Vogel, Steven, 25
Volta, Alessandro, 50
von Berlichingen, Gottfried (Götz, Iron Hand), 27, 28

WalkAide, 40–42, *41*
walking: arm movement in, 151; center of mass, *34*; description of, 31–33; enhancing performance when, 35–36; in force field, 93; leg springs, *34*; Lokomat, 87–88, *89*, 90, 91; retraining for, 90–91; ReWalk, 87, 162; stretch reflexes, *32*
War Machine (graphic novel), *46*, 47, 61
War Machine armor: depiction of, *78*; description of, 77–79; given to Air Force, 152; in *Iron Man 2*, 156–57. *See also* Rhodes, Jim "War Machine"
Warwick, Kevin, 38
Watson, James, 74–75

Wayne, Bruce (character), 119
Weber, Doug, 160–61
weight, muscle output based on, 25
Weyand, Peter, 33
Whiplash. *See* Vanko, Ivan "Whiplash"
whiplash, risks of, 166–67, *168*, 169–72
Williams, David, 109–10, 142
Willis, Thomas, 50
wing chun, 165
Wolf, David, 109, 110
Wolpaw, Jon, 56, 159–60
Wong-Chu (character), xiii, 182

Yinsen, Ho (character), xiii, 128, 147, 151, 152